# CONTEMPORARY STRATEGY

# CONTEMPORARY STRATEGY

## II
## THE NUCLEAR POWERS

*Second Edition*
*Revised and Enlarged*

John Baylis Ken Booth
John Garnett Phil Williams

HM
HOLMES & MEIER
New York London

First published in the United States of America 1987 by
Holmes & Meier Publishers, Inc.
30 Irving Place
New York, N.Y. 10003

*Great Britain:*
Pindar Road, Hoddesdon
Hertfordshire EN11 0HF England



Book design by Gloria Tso

**Library of Congress Cataloging-in-Publication Data**

Contemporary strategy.

Includes indexes.
Contents: 1. Theories and concepts—2. The nuclear powers.
1. Strategy—History—20th century. 2. Military policy—History—20th century. I. Baylis, John.
U162.C65 1987 v. 2 355.4'3 86-22872
ISBN 0-8419-0929-6 (v. 1)
ISBN 0-8419-0930-X (pbk. : v. 1)
ISBN 0-8419-1019-7 (v. 2)
ISBN 0-8419-1020-0 (pbk. : v. 2)

MANUFACTURED IN THE UNITED STATES OF AMERICA

# CONTENTS TO VOLUME I

## THEORIES AND CONCEPTS

# CONTENTS TO VOLUME II

## THE NUCLEAR POWERS

# Preface to the Revised Edition

In 1975 we published an introduction to strategic studies under the title of *Contemporary Strategy: Theories and Policies.* It was not long before it became a standard textbook in both British and American universities and military colleges. The book was reprinted on a number of occasions, and, as a result of continuing demand, we decided to write an updated and expanded edition to take into account the wide-ranging strategic debates and policy changes that have occurred since the mid-1970s.

In the event the second edition proved to be considerably longer than the first. Since we felt that nothing should be left out of such an important and expanding subject, our publishers agreed to produce two books instead of one. This decision was made easier by the natural division within the first edition between the examination of strategic concepts and the examination of the defense policies of the nuclear powers. These two sections were complementary, but they also stood on their own. The same is true of the new edition now published under the titles *Contemporary Strategy Volume I: Theories and Concepts* and *Contemporary Strategy Volume II: The Nuclear Powers.*

In addition to the original chapters on the defense policies of the nuclear powers we have added a new one, entitled "Defense Policy-Making," which provides a general introduction to the subject. It examines those underlying factors that affect defense decision making the world over, such as history, geography, ideology, public opinion, bureaucratic processes, and external events.

The chapters that follow attempt to describe and analyze the defense policies of the United States, the Soviet Union, China, Britain, and France from the Second World War until the mid-1980s. We have not tried to fit the defense policy of each state into a rigid framework, but in each case we have looked at a standard set of issues while leaving the individual authors free to develop those dimensions of policy they have thought distinctive or unique to their own subject. Thus in each chapter some attention is given to the historical context, the relationship between commitments and capabilities, the balance between nuclear and conventional weapons, strategic thinking, the domestic dimension of defense policy, defense decision making, alliances, and future problems and prospects. Students interested in comparative studies could use our approach and findings to identify similarities and contrasts between the five nuclear powers.

# Acknowledgments

The authors would like to express their sincere thanks to Doreen Hamer and Marian Weston, who typed the various drafts of this new edition. Thanks are also due to Colin McInnes, who produced the glossary of strategic terms.

# CONTEMPORARY STRATEGY

## 1

# *Defense Policy-Making**

JOHN GARNETT

### *Policy and Decision*

Since this volume of *Contemporary Strategy* describes and explains the defense policies of some of the world's most important military powers, it seemed worthwhile to devote an introductory chapter to examining and analyzing the sorts of problems that confront all defense planners. Though every state is unique in terms of history, organization, and perspective, in a sense all those who are responsible for government participate in political and administrative activities that are recognizably the same the world over.

Sir David Kelly once made the dry observation that decision making—or is it decision taking?—in the civil service is "casual unreasoning action by ordinary men in positions of extraordinary power." Now although some decisions by every government do sometimes seem to have the quality of inexplicable arbitrariness to which Sir David referred, the assumption underlying this analysis is that, for the most part at least, they reflect a deliberate and calculated response by officials to their situation. In other words, it is assumed that defense decisions involve a reasoned choice from a number of perceived alternatives of that course deemed most likely to promote the state of affairs desired by the decision maker. As such, defense policy is susceptible to rational analysis, and although we may never quite get to the bottom of any particular decision,[1] we can penetrate some of the mystery by identifying the constraints that pushed decision makers toward some choices and away from others. Much of this introductory chapter will deal with the human, organizational, and political pressures that combine to force policymakers in all countries to take actions that, however curious to the outside observer, make sense when located in their proper context.

It may be sensible to begin with a mind-clearing exercise, a little conceptual analysis designed to clarify the meaning of the words we use. In particular, we need to define precisely the term *policy* and to try

*This chapter is a revised and extended version of a paper first delivered at the National Defence College in 1974 and published under the title "Some Constraints on Defence Policymakers" in L. W. Martin, ed., *The Management of Defence* (London: Macmillan, 1976).

to spell out how it differs from *decision*. This is easier said than done because in many people's minds decision making is synonymous with policymaking. Probably no one would dispute that the two concepts are related, but as Heclo has suggested, "The term policy is usually considered to apply to something bigger than particular decisions." In a similar vein Etzioni has argued that policy is a form of generalized decision making "in which whole sets of decisions are considered."[2]

In retrospect at least, policy is revealed by *series* of decisions, and in prospect it is revealed by general statements of purpose. This is not to say that statements of intent are the same as policies. When we talk about a state's intentions we need to distinguish very carefully between what might be called its long-term aspirations, i.e., those goals or objectives that, all things being equal, it would like to achieve, and those it is actually going to act on. World communism may be regarded as a long-term aspiration for the Soviet government, but it is difficult to see what operational implications follow. In terms of practical foreign and defense policy-making in the Kremlin the goal of world communism is so remote from the issues that confront Soviet leaders daily that it is difficult to see how it can have any practical significance. For policymakers struggling with problems of arms control, East-West relations, and the rise of China, and having to cope with failures of technology and agriculture to say nothing of dissident minorities, militant Muslims, and awkward allies in the Warsaw Pact, the goal of world communism must seem completely irrelevant.

Similarly, Western commitments to such long-term objectives as "general and comprehensive disarmament" are a million miles away from the sorts of issues on which governments have to have a policy. The notion of "general and comprehensive disarmament" is simply irrelevant to policymakers struggling with the details of SALT, START, INF, MBFR, etc. It may be the avowed objective of all civilized states, but in terms of policy it has no meaning.

Clearly then, intentions are not the same as aspirations. In the policy-making context, a state's intentions reflect not what it would like to do, but only what it is going to do. The term *policy* embraces both what is intended and what happens in its implementation. The term ought therefore to be "operationally identified not by its goals but by the actual behavior attempting to effect the goals."[3] Policy, therefore, is best thought of not as a series of finite decisions, but as a *flow* or purposive action over time. It incorporates what governments have done and what they are trying to do.

Whatever definition is accepted, it is clear that policies involve decisions—decisions about both what to do and how to do it. Now this distinction may not stand up to rigorous examination, but it is useful because it suggests that there are two kinds of decisions. S. Huntington has made a somewhat similar distinction between what he calls "strategic decisions," i.e., "those made in the categories or currency of international politics"; and "structural decisions," i.e., "those made in the currency of domestic politics," which deal with procurement, allocation, and organization of men, money, and material.[4]

In general, decisions about what to do are policy decisions with a high political input. Thus, the issues of whether to join NATO and the EEC and whether to stay east of Suez involved the British government in policy decisions. Decisions of the how-to-do-it variety are second-order decisions; microdecisions, if you like. They take place *within* the context of policy; they are more technical and in a sense more intellectually manageable. Examples include whether to buy this weapon or that weapon, whether to opt for this mix or that mix of forces. Decisions like these have received the most attention in the defense field.

Today almost everyone is aware of the enormous strides made in developing and exploiting rigorous analytical techniques capable of improving these second-order decisions. In the United States, particularly since McNamara's tenure as secretary of defense, sophisticated management techniques have been introduced to save money, to use resources more economically, and to produce better-designed equipment. The McNamara revolution allocated a central role to the "planning-programming budgeting system," and over the years it has changed the face of defense management in the United States. Operational research, cost-benefit analysis, and systems analysis are now accepted tools of the trade in the defense business. And in many fields their value is proved.

But a word of warning is appropriate. Defense management is not quite the same as defense policy-making. There are a good many first-order decisions to be made in the defense field, and it is doubtful whether the programmatic approach has much to offer by way of making them better. In the euphoria that followed the introduction of these skills there was, perhaps, a tendency to believe that all defense problems could now be solved. M. Desai once commented unfavorably on the tendency to apply the new economic techniques of optimization to highly complicated "messy" situations in which they are quite inappropriate. He cited several incidents from the Pentagon Papers that illustrate the way the Johnson administration tried to reduce to manageable and quantifiable terms Vietnam war situations that were far too complicated for techniques designed fundamentally for relatively simple resource-allocation problems. As Desai said, "A complex phenomenon such as guerrilla war demanding an analysis of many factors, most of them unmeasurable, even unimaginable, by the analyst, has to be reduced to a few variables because the method cannot handle many, and anyway the social scientist believes that the simpler the model the better it is."[5]

This chapter has an explanatory rather than a prescriptive flavor. The aim is to examine not how defense policy-making might be improved, but how particular policies are chosen. This aim is achieved by identifying some of the main constraints and influences that make themselves felt during the decision-making process. Decision makers are seen as acting in "a matrix of internal and external forces that affect their behaviour through pulls and pressures."[6]

## *The Policymaker's Predicament*

Of course, no one would suggest that defense policies are completely *determined* by this matrix of internal and external forces. But there is little doubt that the freedom of decision makers to *form* policy is much less than is sometimes supposed. On appointment, with plenty of energy, and a new-broom mentality, planners sometimes entertain high expectations of what is possible. Indeed, for those who are inexperienced in government, almost *anything* seems possible. It is all a matter of will. On retirement, however, with less energy but more experience, those selfsame planners are often acutely conscious of the constraints that circumscribed their freedom and minimized their achievements. At this time fate seems more important than will. In the rueful words of Arthur Schlesinger, no statesman should expect "to penetrate a process so cunningly compounded not only of necessity but of contingency, fortuity, ignorance, stupidity and chance."[7]

The world of defense decision making is not an infinitely malleable environment, responsive to the thoughts and actions of forceful individuals. Much of what matters is either given or beyond human control. No one can do anything about the facts of geography—the size, the shape, the location of states and continents, and the distribution of resources between them. Nor can anyone significantly affect either the spread of industrialization or the pace and direction of technological innovation. Some things are more or less fixed—power relationships, frontiers, relative economic strengths, etc.—and wise and responsive policymakers must learn to operate *within* those parameters they cannot alter. They still have opportunities for choice, but the *range of choice* is invariably more circumscribed than appears to casual observers.

It seems that the less we know about the problems that confront policymakers, the more freedom we believe they have to move things in the direction of their choice. Only when our knowledge approximates theirs can we fully appreciate the pressures that pushed decision makers in one direction rather than another. Tolstoy made the point like this:

> If we consider a man alone, apart from his relation to everything around him, each action of his seems to us free. But if we see his relation to anything around him, if we see his connection with anything whatever—with a man who speaks to him, a book he reads, the work

> on which he is engaged, even with the air he breathes or the light that falls on the things about him—we see that each of these circumstances has an influence on him and controls at least some side of his activity. And the more we perceive of these influences, the more our conception of the decision maker's freedom diminishes and the more our conception of the necessity that weighs on him increases.[8]

In short, the more we understand the constraints under which human beings operate, the more sympathetic we are likely to be to their failings. In that sense, to understand is to forgive. Without wishing to condone every decision that has ever been taken, I am bound to say that the more I have delved into the detail of particular decisions and policies and the more I have become aware of the pressures on those who take them, the more I have sympathized with the policymaker's predicament.

Perhaps the most obvious and intractable of the external forces that constrain the freedom of action of decision makers are to be found in the harsh realities of power politics. As one writer has expressed it, "Nothing defines the parameters within which a nation can (or must!) act more than its relative power and attributed status in the international order."[9] The facts of international life and the pressure of events over which decision makers have no control inevitably narrow the area of maneuver available to statesmen. For example, as Lord Strang once pointed out, "No British foreign secretary can get away from the fact that Great Britain is a small, densely populated island with wide overseas interests, inescapably dependent upon foreign trade for the maintenance of its relatively luxurious standard of living."[10] British vulnerability in terms of its economic strength, its dependence on trade for sources of food, raw materials, and energy is a powerful constraint on any foreign minister's freedom to act. There are limits to what small, densely populated islands can do in the world, and it is this fact of life that explains a good deal of British policy since the war. Those who cannot mold the international environment, must, inevitably, learn to adjust to it.

Think, for a moment, of the historic decision by Britain after World War II to undertake a firm commitment to the defense of Western Europe via the Dunkirk, Brussels, and North Atlantic treaties. Given the grim international situation—the Soviet threat, a weakened and devastated Europe, an exhausted population, and the possibility of American withdrawal—what else could Britain do but throw its weight into a collective defense arrangement that reversed a long tradition of nonalignment? Statesmanship it may have been, but from another perspective it was determined by the whiplash of events, by the relentless pressures of the international environment, and, in a sense, it was inevitable. Similarly, Britain's withdrawal east of Suez can be interpreted as a response to a changed international situation and economic pressures which, for one reason or another, became irresistible.

In a bipolar world dominated by the two superpowers, the power of medium-sized states is inevitably circumscribed. That is why France has welcomed the decline of bipolarity and the diffusion of power that has accompanied it. In theory at least, a "multipolar" world offers middle-ranking powers greater opportunities for exercising their influence. In practice, the breakup of bipolarity has not gone very far, and even France, exercising all her freedom of maneuver and independence from the superpowers, has found it difficult to escape the limitations of power that constrain all who have slipped out of the first division.

And of course it is not only small or medium-sized powers that find their choices effectively circumscribed by circumstance. The United States, as powerful a state as any and more powerful than most, is almost equally chained by the facts of international life. Stanley Hoffmann has described how after World War II, when the nature of Soviet power and policy and the erosion of British strength became obvious, the United States simply had to fill the power vacuum that was formed. As Hoffman puts it, "The old choice between isolation and involvement is dead; there remains only the purely academic choice between historical sleep and the dangerous life of a great power. The fact that the latter often seems like a nightmare does not make escape into untroubled dreams a genuine alternative."[11]

The Soviet Union is equally fettered by international circumstances. Given the military power of the United States and NATO and the hostility of China on its eastern borders, what realistic alternative is there to massive defense spending? Given the enthusiasm of President Reagan for his Strategic Defense Initiative, what choice do the Russians have about stepping up their own research into the militarization of space? Given the size and destructive capability of the U.S. submarine-launched ballistic missile (SLBM) force, what else could the Russians do but invest heavily in killer submarines and antisubmarine warfare (ASW)? When questions are put like that, Soviet defense policy seems no more than an inevitable result of the forces acting on its policymakers. The notion that they are free to choose radical alternatives is a myth. Hemmed in by world events and the policies of powers over which they exercise little control, Soviet defense planners do what they have to do to safeguard the interests of their country.

## *The Influence of History and Geography*

All states learn by experience. That is why history is such a powerful influence on foreign and defense policy. Given the Russian record of invasions going back to the Middle Ages and a scale of suffering in World War II unprecedented in recent times, who can blame the Russians for their paranoia about security? No wonder they feel the need to overinsure against capitalist encirclement. Russian history provides a persuasive argument for the current military policies of the Soviet Union and the enormous quantities of weapons it has amassed.

Much the same point can be made about French defense policy. How much of recent French suspicion of alliances can be traced to the failures of the international community to save France in both world wars? General Gallois's often-quoted claim that "no nation can be expected to commit suicide for the sake of another" is usually interpreted as a rational response to the nuclear age, but it is also a response to French experience in the twentieth century. "From the French perspective . . . history has taught the twofold lesson that allies can be as unreliable as adversaries are implacable, and that France must have an independent ability to safeguard its interests and security if it wants to avoid being little more than a pawn on the diplomatic chessboard."[12] How much of President de Gaulle's search for grandeur is a straightforward reaction to the ignominy of defeat in two world wars and the painful loss of empire in North Africa and Indochina? To what extent is the assertiveness of French nationalism and independence a corollary of previous weakness? It is impossible to answer these questions with any confidence, but simple psychology suggests that they are worth asking. No state can escape its history.

In a similar vein, one cannot help wondering how far modern China's defense policy is a reflection of that country's historical experience. How much do the humiliations experienced by China at the hands of Western powers during the nineteenth century color present policies toward the West? China's humiliation must have been felt all the more keenly following its traditional self-perceived cultural superiority. Such factors are difficult to quantify, but common sense suggests that the "century of humiliation" following the Opium War and the various "unequal treaties" imposed upon China have left contemporary Chinese statesmen with a backdrop of suspicion and resentment that is difficult to overcome. China has, of course, abrogated the "unequal treaties," which robbed her of so much territory, but it seems likely that current sensitivity over the lost territories and a determination to become a great power have historical roots in Western exploitation.

One can also see the influence of history in the Chinese obsession with guerrilla warfare. The tradition of banditry and peasant rebellion, stretching far back in Chinese history, was powerfully reinforced by the success of Mao Tse-tung in overthrowing the Kuomintang in a series of protracted and brilliant revolutionary campaigns. The success of the Communist revolution in 1949 helps explain why, even in a post–cultural revolution society, the idea of people's war still dominates the thinking of many Chinese defense planners. The relevance of nuclear deterrence, modern weapons, and air power is much less immediately obvious.

If history provides one key to unlocking the mysteries of defense policy, then geography surely provides another. How much of Soviet emphasis on conventional power can be explained by the sheer size of the Soviet Union and the vulnerability of its frontiers? The absence of obvious natural boundaries between Russia and most of its neighbors has been a permanent source of concern for successive generations of Russian statesmen. Guarding thousands of miles of frontier remote from centers of decision and often in areas where communications are inadequate is a daunting prospect and goes a long way toward explaining the Russian insistence on massive conventional forces.

Similarly, it can be argued that the enormous extent to which the Chinese have relied on the concept of people's war is partly explicable in terms of Chinese geography. With millions of square miles of territory and virtually unlimited human resources, it makes sense for the Chinese to drown their enemies in people's war. Any army contemplating the invasion or conquest of China runs the risk of being swallowed up by the vastness of Chinese territory. Overextended and dispersed attacking forces would quickly lose their momentum and would eventually fall victim, first to guerrilla tactics and then to more conventional operations and counterattacks. Various observers have pointed out that China's size and terrain make it an ideal country for exploiting guerrilla tactics. In the words of Defense Minister Xu Xiangqian, "China must continue to make the most of its vast territory and large population. . . . If any enemy dares to invade China, the main forces of the country and local corps will annihilate large numbers of enemy troops. The militia will be mobilized to look for any opening to hit the enemy. Everyone will be a soldier, every village a fortress and every place a battlefield. No enemy can stand the attrition. . . ."[13]

Gen. Pierre Gallois has commented on the powerful part played by the geographical factor in determining French defense policy. "France has a specific role to play in the nuclear age which is defined by her continental position and the limited resources at her disposal to defend the security of that position." Sabrosky has pointed out how French defense planners are constrained by the vulnerability of French territory to attack from the East. Nevertheless, this vulnerability is moderated by the buffer of Germany, which stands "between France and the most likely source of a military threat to France." The existence of this buffer zone "allows French planners to formulate a strategy, articulate a doctrine, and prioritize their forces in a way that would simply not be feasible if France were located in the same situation as, for example, the Federal Republic of Germany." Sabrosky also points out that the hard geographical fact that France has insufficient quantitites of at least one key strategic resource—oil—means that "France is a Continental power that must act as both a 'Continental' and a 'maritime' power to secure both its position and its resources."[14] French statesmen of whatever political persuasion cannot avoid taking these geographical facts into account when they formulate French defense policy.

## *Commitments, Obligations, and Domestic Environment*

Decision makers who are constrained by international circumstances, by their country's historical experience, by the facts of its geography, and by contemporary events, are also constrained by their commitments and obligations and the web of relationship they inherit. In this sense they are constrained by the decisions of their predecessors. Any alliance inhibits the freedom of movement of every member, even the most powerful. Every treaty obligation does the same. Even membership in the United Nations carries obligations. Some of these commitments may be purely residual, relics of a bygone era; but however inconvenient they may be, they are legally binding, and their existence helps to determine the size, the shape, and posture of a nation's defense effort.

A revealing example of this is provided by the recent Falklands campaign. As a result of an almost forgotten obligation the British government was forced to fight a bloody war and to commit, for the foreseeable future, enormous resources to defend islands that, on any common sense reckoning of the national interest, are almost totally worthless. The "fortress Falkland's" policy is a classic case of the tail wagging the dog. Prime Minister Thatcher has been trapped into accepting such a policy by circumstances quite beyond her control. Once Argentinian troops had invaded, the politically acceptable policy options available to the British government were reduced to one—recapturing the islands.

And successive governments may find themselves equally boxed in. When lives have been lost and vast expenditure incurred, commitment is unavoidably strengthened. In a sense, therefore, the British government now has even less room to maneuver than it had before the task force landed. Handing the Falklands to Argentina with some sort of "lease back" guarantee may have once been a plausible option—and one day it may become so again—but it is not a realistic policy now. The British people have sacrificed too much too recently to stand for it, and even if they did, the Conservative government could not contemplate it. To recapture the Falklands and then to promptly hand them over to the aggressor is asking too much of any government—even though, in hardheaded national interest terms, it might make sense.

Policymakers are constrained by the facts of international life and the web of relationships and commitments in which their country is enmeshed. But because they are part and parcel of it, they are even more constrained by their domestic environment. They share domestic values, and they are constantly exposed to the pressure of both organized and unorganized groups at home.

Of all the domestic constraints that impinge on policymakers, the most insistent is the financial one. Just as all of us, as individuals, find our life-styles cramped by the inadequacy of our economic resources, so defense planners are persistently constrained by the hard facts of economic life. "The bottom line" sets the parameters within which defense planning has to take place.

Since the provision of national security is only one of the activities for which governments are responsible, ministries of defense have to compete with other government departments for a share of the national cake. At any one moment, the amount of money devoted to defense reflects the priority the government has allocated to it in competition with education, health, social security, transport, agriculture, etc. In a sense, therefore, defense expenditure is determined not by *economic* constraints, but by *political* decisions—decisions that reflect the values and priorities of the electorate and the government. Some governments—those to the right of the political spectrum, for example—may prefer to buy more security and less welfare. Others, particularly those on the left of the political spectrum, will probably favor welfare at the expense of security. Either way, the amount of money allocated to defense reflects political choice. "If expenditure is cut back this is not because this is in some absolute sense necessary, not because the nation could not afford the previously planned effort, but because certain choices are made."[15]

Currently, Britain is spending just under 5 percent of its GNP on defense. This percentage is somewhat higher than those of its European allies and comparable to that of the United States. But it is much lower than that of the Soviet Union. Although defense spending is often expressed as a percentage of GNP, such statistics may prove very misleading and unhelpful. Ultimately, what matters is the actual amount of money expended. For a country with a very high GNP even a low percentage spent on defense may represent a good deal of money.

Obviously, no defense budget is set in concrete, and expenditures will go up or down according to the government's changing perceptions of the various competing claims on the public purse. But once expenditure is fixed for a benchmark program stretching over a number of years into the future, it sets an absolute limit on the amount of money that can be spent. The fact that all government departments have to operate within their existing and projected budgets is a powerful constraint on defense planning. What is planned has to be affordable.

In practice the constraint is even tighter. Even when it has been allocated, money for defense is not "free money," available to defense policymakers to spend as they please. Most of it is already committed to a particular pattern of defense spending—in much the same way that most of the average person's monthly salary is already committed to a fixed pattern of mortgages, school fees, insurance policies, rent, etc. No government starts with a clean slate. It *inherits* a defense policy, and that means, as D. Greenwood puts it, that "you start from where you are."[16] Probably the most important explanation of the size, shape, and equipment of any state's defense effort is *habit*. Though they are able to make changes at the margin, existing programs reflecting unalterable commitments, sometimes for years ahead, severely restrict the defense planner's freedom to maneuver.

In France, for example, recent governments have made fairly determined efforts to redress the balance between nuclear and conventional forces. For years the *force de dissuasion* (The French nuclear deterrent) has drained defense resources away from conventional forces. The *loi de programmation 1977–82* sought to shift defense priorities by placing greater emphasis on reorganizing the army and upgrading conventional general-purpose forces. But it is difficult to escape the conclusion that despite its declaratory policy the French government has found it impossible to implement this reassessment. The critical role of nuclear weapons in French policy requires the continuing expenditure of money to improve them. The pattern cannot easily be changed, and the 1984–88 program admitted as much by proposing cuts in conventional forces so that more could be spent on the French deterrent.

## *The Constraint of Value Systems*

Some domestic constraints are scarcely felt at all by decision makers. Those who are reared in a particular value system rarely feel constrained by it even though it determines everything they do. So those restrictions on decision-making freedom that arise out of an unconscious compliance with a particular political style, for example, may not impinge on the decision maker at all. The British reluctance to think far ahead, to discuss abstract ideas, to see questions in terms of principles, and to discuss hypothetical situations are all characteristics of what is sometimes unfairly called the philosophy of "muddling through." This pragmatic tradition of policy-making comes so naturally to civil servants reared in it that they are scarcely conscious of it as a constraint at all; but few would dispute the effect it has on the way in which the civil service sees problems and handles them.

Though few would dispute their influence, the values that underpin the perceptions of Western bureaucrats are rarely made explicit. But in countries with clearly articulated political ideologies, the values that shape policy may be more easily identified—at least by outsiders. Russian decision makers, reared in the ideology of communism, have inevitably absorbed certain ideas about human society, international politics, and the historical process. These ideas have an impact on defense policy-making in the Soviet Union just as they affect almost every other aspect of Soviet politics and life.

Those who believe there is an inherent antagonism between capitalist and Communist states are likely to put a high premium on permanent vigilance and are likely to be skeptical about the possibilities of détente and arms control. Those who believe in the inevitability of ultimate victory are likely to look at problems of war fighting in a slightly more optimistic way than those who do not share this comforting assurance. Those who believe in revolutionary change are likely to regard political instability in the Third World as an opportunity rather than a threat. It is difficult to pinpoint the precise policy implications of these beliefs, but common sense suggests that they must form an influential, indeed an inescapable, backdrop against which all Russian defense policy is formulated.

A similar point can be made about Chinese defense policy. As J. Baylis points out (see chapter 4), in China there is permanent tension between the ideological influence of communism and Mao's doctrines and the pragmatic requirements of modernization and professionalism in the armed services. Though the death of Mao and the demise of the "Gang of Four" may have tipped the balance in favor of the latter, the constraint of ideology is ever-present. Given its history and ideological roots, the People's Liberation Army can never be *entirely* professional. Similarly, Chinese foreign policy can never be *entirely* pragmatic. Ideas about world revolution and wars of liberation may not have had much practical impact, but they are part of the intellectual justification for Chinese foreign policy, and their influence cannot be ignored.

Just as a policymaker may be unaware of the constraints imposed upon him by the value system and style within which he operates, so he may be unaware of the way in which his own personality inevitably shapes the decisions he takes. But though he may be unaware of it, it is clear that the personal characteristics of the decision maker and the prism through which he looks at the world have something to do with the way in which he perceives problems and tries to solve them. P. Darby makes the point that between 1956 and 1959, many of the problems that Duncan Sandys, the British minister of defense, faced in dealing with the services can be explained by his own idiosyncrasies. "Service grievances perhaps turned more on the personality of the minister and his method of taking decisions than on the doctrinal content of his policy."[17] Apparently Sandys had the unfortunate habit of "disregarding professional advice," a penchant for "asking fundamental questions," and an unfortunate tendency to "treat the Chiefs of Staff like schoolboys."

Though policymakers may be quite unconscious of some of the constraints under which they operate, they are very much aware of others. Public opinion, for example, is a domestic constraint that few government departments can afford to ignore. Of course, on most specific defense issues the public has no opinion, and, indeed, there is plenty of evidence of popular apathy across the entire foreign and defense policy field. In almost every British election since the war it has been questions of economic and social policy rather than defense that have dominated the political manifestos of the major political parties. But there is an important negative sense in which this lack of interest or enthusiasm for defense is an important constraint on governmental behavior. As Franklin D. Roosevelt once said, "I cannot go any faster than the people will let me," and so long as the British public allocates such a low priority to defense, no British government can pursue anything but a low-profile defense policy. In particular, the defense budget could not easily be raised during a period in which many people are reluctant to maintain even present levels of expenditure.

It seems probable that the lack of public interest in defense that has characterized postwar British politics is coming to an end. There are a number of reasons for this. First, because the essentially biparty defense policy that has prevailed during the period has now quite clearly broken down. For the first time, there are now major differences between Conservative and Labour defense policies—particularly in the nuclear field. In the years to come, defense is likely to be a major election issue and will therefore enjoy a high profile in domestic politics. Public opinion may very well become as polarized as party political opinion, and even if it does not, public interest in defense is bound to grow.

The second reason for thinking that public opinion will, in the future, become a more powerful constraint on defense policy formulation is that the various pressure groups that, taken together, loosely constitute the "peace movement" are determined to keep defense issues at the forefront of public consciousness. Though the Campaign for Nuclear Disarmament (CND) movement may have failed in its primary purpose of persuading the British government to abandon its nuclear deterrent and to remove cruise missiles from British soil, the movement has already done a great deal to heighten public awareness of strategic and arms control issues.

The third reason the general public seems more likely than ever to interest itself in defense matters is an economic one. These days, militarily effective defense policies are inordinately expensive. Annual British defense expenditure is now running in excess of £18 billion. Apart from social security, defense is now the second biggest item of public expenditure—having, in recent years, overtaken expenditure on health and education. In a tough economic climate in which welfare services are being seriously eroded, it is inevitable that the public will question the priority currently given to defense.

Like it or not, defense has become, and is likely to remain, a high-profile activity. Governments can no longer afford to neglect public opinions about it. This is true not just of Britain, but of Western countries generally. A persuasive example of the power of public opinion in European countries is provided by the reluctant acquiescence of both the Dutch and Belgian governments in accepting the deployment of cruise missiles in their territory. Both governments have been extremely sensitive to public opposition to the 1979 NATO Council decision to modernize NATO's theater nuclear capability.

But in France, on nuclear matters at least, there has developed an unusual public consensus between Left and Right on the continuing need for a French deterrent. Communists and Gaullists alike support the *force de dissuasion,* and there is no real equivalent to the popular nuclear protest movements that thrive in other Western European countries. Pacifism is peculiarly unpopular in France. The reasons for this unusual consensus between almost all parts of the political spectrum are difficult to determine, but it seems that the French public is basically apathetic and resigned to the status quo on defense matters. What this means is that all political parties, even those on the Left, that have a serious interest in attaining power have had to trim their policies to a grudging support of nuclear deterrence. Whether this public consensus will continue into the late 1980s is problematic. Recent studies suggest that it may be beginning to crumble.[18]

In the United States presidential difficulties in persuading the Senate to fund the MX missile program provide another example of the power of public opinion to frustrate and amend government policy. The influence of the environmental lobby in killing the Multiple Protective Structure (MPS) "basing mode" program for the protection of MX missiles is also an impressive example of how public opinion can be brought to bear on a precise issue.

Most political scientists are aware of the significance of public opinion to government policy-making, but many are skeptical of its value. There is widespread agreement that it is inchoate, fickle, unreasonable, inconsistent, and arbitary. Theodore Sorensen summed it up when he said, "It rarely considers the needs of the next generation or the history of the last. It is frequently hampered by myths and misinformation, by stereotypes and shibboleths, and by an innate resistance to innovation. It is usually slow to form, promiscuous and perfidious in its affection and always difficult to distinguish."[19]

But for all its faults, no democratic government can afford to ignore public opinion for long, and any defense policy that cannot be accommodated within it is purely academic. If only in this negative way, public opinion is a significant constraint on the formulation of defense policy. But there are many occasions when public opinion, expressing itself through highly organized pressure groups, has had much more positive influence over decision makers.

In the United States in particular, a large variety of interest groups seek to influence the behavior of government in the field of defense. Environmentalists and antiwar groups like the National Committee for a Sane Nuclear Policy, the Council for a Livable World, the Center for Defense Information and the Nuclear Weapons Freeze Campaign lobby the federal government and marshal public opinion on defense matters.

But the most powerful pressure groups in the defense field come from within the military-industrial complex. American armed forces personnel number close to two million, and, led by sophisticated service chiefs, they exercise enormous influence in Washington. So too do those industries that manufacture the hardware on which American strength is built. Large sums of money are spent on weapons procurement each year. Most defense contracts go to a relatively small number of major corporations—Lockheed, General Dynamics, General Electric, Boeing, McDonnell Douglas, etc. Those companies work closely with the Department of Defense and with those senators and congressmen whose constituents benefit from the work generated by defense orders. Inevitably their spokesmen exercise considerable influence over the direction of the defense effort. In chapter 2, Phil Williams discusses how the military-industrial complex has brought pressure to bear on the B-1 bomber project and has kept it alive for years.

It is probably true that no aspect of American defense policy is entirely free from the involvement of one interest group or another. By fighting their respective corners, pressure groups contribute enormously to the diversity of ideas that are fed into the government machine. They have a crucially important role in the American political system. But in the defense field, concern is frequently expressed about the extent to which these private organizations have come to influence government policy. A bogeyman is conjured up of a vast, unified military-industrial complex, exercising disproportionate power in a conspiratorial, even corrupt, fashion, and quite beyond democratic control.

In a country where one job in ten and twenty-five scientists and engineers out of every hundred depend directly or indirectly on defense spending, the problem is one of sheer size and public accountability. In the words of a recent columnist in the *London Times,* "Knit together by mutual self interest, by shared beliefs and by patriotism, the Pentagon, the armed forces, giant corporations, political leaders and legions of defense scientists, scholars and salesmen form an institution that touches every corner of the country."[20] With such enormous spending power at its disposal, this lobby can captivate trade unionists, government officials, and congressmen alike. No administration can remain immune to pressures that are insinuated into every corner of defense activity.

Though more than thirty years have passed since President Eisenhower first alerted his fellow citizens to the dangers inherent in such an overwhelming concentration of power, little has been done to control it effectively. The military-industrial complex remains a source of anxiety to all who are interested in democratic values.

In Communist countries, of course, the situation is quite different. There are no Soviet equivalents of the various American pressure groups. Special pleading is done, not through organs of public opinion or interest groups, but through the mechanisms of bureaucratic politics within the structure of government itself.

## *Rationality and Defense Policy*

It might be tempting to regard policies and decisions in the defense field as the result of considered reflection by omniscient and reasonable people pursuing clearly articulated and logically compatible goals. Roger Hilsman has commented that "we would like to think not only that policymaking is a conscious and deliberate act, one of analyzing problems and systematically examining grand alternatives in all their implications, but also that the alternative chosen is aimed at achieving overarching ends that serve a high moral purpose." This view that decisions are taken by unitary, high-minded individuals who are seeking to maximize the national interest is a travesty of reality, but it is implicit in much writing in the defense and foreign policy field. It is also an unconscious paradigm in the minds of ordinary men and women who, when confronted with some inexplicable piece of government behavior, shrug their shoulders with the words, "I suppose they know what they are doing." Graham Allison has caricatured this notion of decision making as the "rational policy model," which draws an analogy between a government and a reasonable man, and assumes that government behavior can be explained in terms of the means which a rational man, confronted with the same situation, would adopt to achieve his goals. This model of analysis can produce plausible but by no means adequate accounts of international behavior. Thus, Soviet missile policy in the Cuban missile crisis is usually explained as being motivated chiefly by "the Soviet leaders' desire to overcome . . . the existing large margin of U.S. strategic superiority,"[21] and the American quarantine response is explained as the most sensible and reasonable choice from among the options considered by the Executive Committee of the National Security Council (Ex-Com. Committee). The same rational policy model assumptions are built into most conventional interpretations of arms races as action-reaction processes. For example, Soviet-American rearmament programs during the 1950s are usually explained on these lines:

> When, in 1954, the Soviets demonstrated the capacity to produce medium and heavy jet bombers, this led to fears in the United States of a "bomber gap" and an accelerated B-52 program. In 1957, the Soviet announcement of an ICBM capacity, plus the orbiting Sputnik, led to American fears of a "missile gap" and increase in U.S. missile expenditure.

Those who favor this kind of analysis—and to some extent all of us do—appear to think that an action in international politics is rather like a move in a complicated game of chess, to be explained entirely in terms of a desire to win (promote the national interest), and as a rational response to an opponent's moves. Explaining policy is simply a matter of showing how, given their national interests and a particular situation, governments could rationally choose such policies.

This way of thinking has a lot in common with what Charles Lindblom has described as the "synoptic approach" to problem solving, according to which decision makers first sort out and order their objectives and values, then survey and evaluate the various means of achieving them before finally choosing those means best calculated to maximize the objectives attained.[22]

What is wrong with this approach—which, as one writer has pointed out, has "prompted a Panglossian literature on Crisis Management"[23]—is that it fails to recognize that the decisions of governments are not made by rational individuals but by complex, interrelated organizations. Much of what follows is about the way in which these government organizations constrain policy-making in defense.

The flavor of policy-making activity in these vast bureaucratic organizations has been brilliantly described by Allison:

> Hundreds of issues compete for players' attention every day. Each player is forced to fix upon his issues for that day, fight them on their own terms, and rush on to the next. Thus the character of emerging issues and the pace at which the game is played converge to yield government "decisions" and "actions" as collages. Choices by one player, outcomes of minor games, outcomes of central games, and "foul-ups"—these pieces when stuck to the same canvas, constitute government behaviour to an issue.[24]

Another illuminating metaphor in the context of organizational decision making is that of a canoeist on a fast-flowing river. The canoeist's choices about where he travels are severely circumscribed, and his main activity is not so much deciding where to go, but how to survive the rocks and currents that threaten his safe passage.

In short, decision making, like life, is one damn thing after another, and the term *policy,* taken in the normal sense of implying rational goal-oriented activity, is perhaps not an entirely appropriate description of the curious mixture of chance and choice that we are talking about. As Hilsman explained it, "Very often policy is the sum of congeries of separate or only vaguely related actions. On other occasions it is an uneasy, even internally inconsistent compromise among competing goals or an incompatible mixture of alternative means for achieving a single goal."[25]

In other words, instead of dealing with a rather highfalutin activity in which powerful men decisively change the course of human affairs, we are discussing the much-maligned philosophy of "muddling through," of fumbling forward in ministeps without much knowledge or confidence, but in the vague hope that we are moving in the right direction. Charles Lindblom has analyzed the process of disjoined or incremental problem solving, according to which policymakers confronted with a problem simply devise tactics that involve only very small or incremental changes from their existing policy. "In short, policymakers and analysts take as their starting point not the whole range of hypothetical possibilities, but only the here and now in which we live, and then move on to consider how alternatives might be made at the margin."[26]

In the late 1970s within the Ministry of Defence the question of whether Britain should remain a nuclear power may or may not have been raised, but the question of whether to put multiple, independently targeted re-entry vehicles (MIRVs) on Polaris or go for Poseidon or Trident was certainly discussed. Now it may be argued that choosing between Polaris, Poseidon, and Trident assumes a prior decision to retain the nuclear deterrent; but if the problem did not come up in that form, it is unfair to see it in that way. An incremental decision to replace one system with another was required. Perhaps that is what Chapman Pincher meant when he wrote, "a *technical decision* has been taken during the life of a Labour government to replace the few existing Polaris submarines with new submarines carrying much improved Polaris-type missiles [italics mine]." Of course, that incremental, technical decision, by foreclosing other alternatives, inevitably implied a major policy decision. As Brzezinski and Huntington have suggested, even "the grand choices are often made more by accretion than by selection."[27] The decision to move to the next generation of deterrent weapons effectively settled the issue of whether Britain will remain a nuclear power, but if those were not the terms in which it is considered, then that is not the decision that was taken.

This interpretation of the decision-making process once and for all destroys Lindblom's synoptic approach and Allison's rational policy model with their inbuilt assumption that policy-making is a purely rational phenomenon. Decisions are now seen, not as "discrete acts with recognizable beginnings and sharp, decisive endings"[28] but as outcomes of a much messier process. In particular, policy is not made by isolated and rational leaders and imposed downward on the organizations for which they are responsible, but as the outputs of organizations that finally receive a stamp of approval by those who lead them.

Now the fact that decisions originate within organizations means two things. First, that the process by which they are reached is to some extent a *political,* rather than a *reasoning* process. That is to say, the interplay of power and interest on the part of various individuals and sections is at least as important a determinant of policy as the arguments about the merits of particular courses of action. Decisions emerge, not so much as a result of argument and reflection about the problem in hand, but as a result of organizational strife between rival departments, sections, and individuals. The point is that in a given situation, organizations and the individuals within them, though not unaware of the national interest, are in fact pursuing more complicated goals in which organizational prosperity, survival, and individual promotion may figure prominently. Of course, organizational success is not necessarily incompatible with the national interest. What is good for General Motors may indeed be good for the United States; but there is no inevitability about it.

It is difficult to estimate the importance of bureaucratic politics on defense decision making, but even a casual perusal of the literature suggests that it has played a significant part in a variety of important decisions. The history of interservice relationships both in Britain and the United States is full of clashes in which the national interest seemed to get lost in a more parochial struggle for organizational survival. In Britain, for example, the issue of whether or not to abandon the aircraft carrier may be regarded as a struggle between the Navy and the Royal Air Force (RAF) for limited funds. And whatever the merits of their respective arguments, few would dispute the bitterness with which each side pursued its interests. In the event, as Sir Frank Hopkins ruefully revealed, the RAF was better as the politicking than its rivals. In a very revealing passage he described the way in which the RAF conspired to show that the carrier force could be effectively replaced by shore-based aircraft.

> Many devices had to be resorted to . . . such as assuming the existence of bases which were not there and never likely to be; crediting the F-111 with a performance in which even its most ardent supporter could scarcely believe, and in the event never materialized; assuming almost superhuman achievements in logistic support by the RAF; assuming overflying rights of countries in Europe, Africa and Asia which were unlikely to be allowed in the event; and even, in one study, moving Australia 600 miles to the North-West in order to bring certain targets within the already elastic radius of action of the F-111.[29]

A detailed analysis of a similar kind of in-fighting in the United States is to be found in P. Y. Hammond, "Super Carriers and B-36 Bombers: Appropriations, Strategy and Politics," and Vincent Davis's book *The Admirals' Lobby,* which is a very revealing analysis of the naval establishment as a pressure group. Hammond clearly explains how some of the major strategic weapons purchased by the United States after World War II were brought not so much for strategic reasons but as a result of organizational pressures from within the services. Davis pertinently points out that during World War II, in pursuing their interests, "Air Force officers . . . sometimes seemed incapable of distinguishing between their dedication to country and their dedication to proving the validity of their Douhet-Mitchell strategic ideology." Later on, in the late

1950s, when it was the air force's turn to feel threatened, this time by the advent of the missile, the same organizational parochialism manifested itself. "In the controversy over the proposed B-70 bomber, the willingness of the Air Force men to relabel this plane a reconnaisance aircraft after it had been denied to them as a bomber seemed to say: We feel that we must have this airplane regardless of what it is called and what it may be used for."[30] More recently, interservice rivalry has affected decisions about ground- and air-launched cruise missiles, the MX program, new aircraft carriers, and the B-1 bomber.

We know very little about defense policy-making in the Soviet Union, but there is little evidence to suggest that the Russians are less bureaucratic than we are. Since the fall of Khrushchev, the USSR has been in the hands of a committee rather than a single individual. In the words of David Holloway, "power has been diffused at the center,"[31] and, inevitably this has created greater opportunities for politicking. It seems likely that the same rivalries that beset the defense decision-making process in Western countries will affect Soviet policies equally deeply.

But we cannot be sure of this. At least one observer has queried the conventional wisdom that institutional interests dominate the top echelons of the Soviet power structure in much the same way as they do in Western states. Harriet Fast Scott has warned us of the dangers of "mirror imaging." She has pointed out that power is still very much more centralized in the Soviet Union than it is in Western countries, and she has suggested that struggles within the Kremlin are largely the result of individuals or groups seeking power for power's sake. They are not fights about "guns or butter," heavy industry or agriculture.[32]

Another reason for thinking that bureaucratic politics may be somewhat less important in Soviet defense decision making is that in Russia there is much more general agreement about the priority to be given to defense matters. There may be interservice rivalry and clashes of interest within the defense establishment, but on the overall importance of defense in government priorities there is a good deal of consensus both within and outside government circles. Given the frequency with which Russia has been invaded and the suffering she has experienced, this is not surprising. But it does mean that defense policy-making in the Soviet Union is not subjected to quite the same pressures as those that exist in countries with less tragic histories.

Our knowledge of Chinese defense decision making is, if anything, even more fragmentary than that of Kremlin politics. But, to quote one authority, "All the available evidence indicates that the policymaking process in China is characterized by conflict, coalition, and compromise among competing bureaucratic interest groups. Unfortunately, however, the evidence is not detailed enough to indicate clearly which policies each group prefers."[33]

The fact that decisions are often made by bureaucratic organizations means that policies inevitably reflect the habits, prejudices and perspectives of those organizations. So pervasive is this limitation that it affects not only the policies that emerge, but even the way in which problems are formulated in the first place. Anyone interested in explaining decisions in the field of British defense policy therefore is forced to consider the way in which the Ministry of Defence procedures by which they are reached rubbed off on them. Clearly, any large organization needs set procedures and detailed rules in order to work at all, and yet these administrative conventions, which are designed to promote efficiency by reducing work to a routine, inevitably inculcate many of the vices of bureaucracy in the minds of those who operate them. To some extent, rigidity of mind, or "hardening of the categories," as one professor so aptly called it, lack of imagination, absence of creativity, and operating by the book instead of thinking for oneself are all vices inherent in bureaucracies. The effect they have on policy-making cannot be overstated even if it cannot be nicely calculated. British readers familiar with the TV program "Yes Minister" do not need reminding of the power of the British civil service, but Sabrosky has drawn our attention to the equally impressive influence of the French civil service. He has alerted us to its power to delay or kill any initiatives it opposes. He says "the fact that the French civil service includes a substantial number of middle- and upper-level personnel appointed under Gaullist administrations certainly inhibits any major departures from Gaullist policies and objectives."[34]

Following the cancellation of the TSR-2 aircraft in Britain in the mid-1960s, there was speculation that the decision may have been influenced by the Ministry of Aviation's accounting procedures. That is, this airplane may have looked so expensive in part because a good deal of general R & D expenditure that ought to have been costed separately, was, for the sake of convenience, lumped in with the TSR-2. Much of the electronics research was relevant to both the Phantom and the Harrier, and it was certainly not wasted. As the authors of a Royal United Service Institution (RUSI) publication on the subject pointed out, "A significant proportion of the estimated £280 million research and development bill for the TSR-2 may have been assigned to it as a result of fairly arbitrary Ministry of Aviation accounting conventions."[35] If this is the case, it provides a classic example of how "standard operating procedures" may distort a problem and significantly affect decisions relating to it.

An equally revealing example of how policy can be affected, if not determined, by organizational procedures is recounted by Admiral P. Gretton. Apparently during World War II the mine barrage laid between the Orkneys and Scotland was a waste of effort in that it made no discernible difference to the passage of enemy surface ships or U-boats. What is so depressing, however, is that "we had laid a similar type of barrier between Norway and Scotland in 1917–18 with similarly ineffective results. But due to the misuse of the rules of secrecy, the planners of the World War II minefield were informed only of the war propaganda magnified claims of success, and the true figures—three U-boat losses only—were kept in the top secret safe." As K. Booth has commented, the Rosengarten minefield "is an instructive example of the way in which an organizational standard operating procedure (classification) can severely affect policy."[36]

A more recent example of the same phenomenon is described by Allison in his very influential analysis of the Cuban missile crisis.[37] When the United States Air Force was asked by the Ex-Com. Committee to spell out the details and implications of the "surgical" air strike option, it started from an existing contingency plan for massive action and came up with frightening details of multiple sorties, likely casualties, and collateral damage. The precise, "surgical" strike was not considered for a whole week because, says Allison, according to U.S. manuals, Soviet intermediate-range ballistic missiles (MRBMs) in Cuba were classed as "mobile" and therefore required extensive bombing. Only when the Soviet missiles were reclassified as "movable" was progress made and the surgical air strike given proper consideration.

The gist of the argument so far has been to suggest that policies and decisions "are not the product of expert planners rationally determining the actions necessary to achieve desired goals. Rather they are the product of controversy, negotiation and bargaining among different groups with different interests and perspectives." As such most decisions taken in the real world can best be described as "extrarational" or "metarational."[38]

Although our focus has been on the way in which the government bureaucracy may modify policy, it is worth pointing out that other organizations can also constrain the freedom of policymakers. In Britain one of the most important of these is the political party of the government in power.

In any democratic political party, leaders walk a tightrope between leading and following; those who emphasize the former at the expense of the latter run the risk of finding their power base dangerously eroded. Government authority rests with the leadership only "so long as they retain the confidence of their political parties." Leaders, therefore, are constantly engaged in consensus politics, in devising policies with sufficient appeal to carry the rank-and-file party members. Inevitably, though consensus building is sometimes costly in terms of compromising the policy's content, it is a prerequisite of effective policy. For that very reason the relationship between a government and the party it represents is best thought of as a continual dialogue, an uneasy partnership between those who have to govern in an intractable world and their ideological supporters, who expect them to implement political manifestos. Very rarely does the dialogue break down completely, though it is frequently strained. In general terms, as R. Butt puts it, "The main lines of government policy are determined by the government but with some shading at the edges to maintain backbench [rank-and-file legislative] consensus."[39] It is this "shading at the edges" that must be explored.

A number of writers have described the way in which intraparty wrangling has constrained the defense policies of successive Labour leaders in Britain. On numerous occasions since the Second World War, considerations of party unity have pressed leaders into policies not entirely to their liking. Whether the issue was German rearmament, the multilateral force (MLF) proposal, overseas commitments, or defense expenditure, the influence of the left wing compelled successive Labour leaders, whether in opposition or in government, to trim their policies, both in form and substance, to accommodate the views of radicals who had no time for responsible realpolitik.

On the question of nuclear deterrence, the policies of the Labour party have, until very recently, been consistently ambiguous partly because the leadership has had to be sensitive to criticism from the pacifist, unilateralist element in the rank and file. On the question of whether Britain should reduce its defense commitments both east and west of Suez, there is plenty of evidence that the Labour leadership was prepared to listen to backbench opinion. When, in the 1967 Commons debate on the defense white paper there were sixty-two abstentions, Prime Minister Wilson admitted that "I now see more time and care should have been given to anticipating the trouble which broke when the vote was called." Thereafter he took steps to become more responsive to the party.

One writer has described how, to reconcile left-wing views with the realities of responsible government and to give itself freedom to maneuver, the leadership was forced "by stealth and subterfuge . . . to draw the substance out of the policy while disguising this with Socialist rhetoric, vague illusions, shadowy formulas and partisan sniping." Antony Verrier also commented that the weak position of the leadership within the party on defense issues led to the development of an "intriguing bag of generalities" rather than a defense policy.[40] These examples all indicate the constraining influence of intraparty debate on the formulation of defense policy. The need for consensus building within the party is an important constraint on the defense policies of all British governments, but it is probably fair to say that it is most striking in the Labour party.

Indeed, for Labour, the problem has gotten worse in recent years because constituency party workers—often activists on the Left—have repeatedly sought to make party resolutions passed by the Labour Party Conference binding on Labour members of Parliament. If they are ever completely successful in this attempt to make MPs responsible to the grass roots from which their power is derived, the freedom of future Labour governments to determine defense policy will be even more circumscribed by party pressures than it has been historically.

In most Western-style democracies, governments have to be sensitive to the claims of their party supporters; but in the United States, though the president needs his party's support in order to get elected, thereafter he can afford to neglect it somewhat. Of course, he still needs support in both the House of Representatives and the Senate if his policies are to enjoy the financial backing necessary for their implementation. But that support may very well come from members of *both* political parties. The point to remember is that in the United States party cohesion is much less than in Britain. Parties are loose, umbrella coalitions, and party discipline is slack. Hence, in Congress, it is not at all unusual for party members, perhaps taking their cues from the relevant committee, to vote against their leaders. And if he can muster a majority, the president may not care where his support has come from.

Traditionally there has been a good deal of bipartisanship in American foreign and defense policy. "Politics stops at the water's edge" is an adage that has worked well for many presidents; but from the 1960s onward, successive administrations have run into increasing opposition in Congress to particular foreign policy initiatives and rising defense appropriations. During the 1970s Congress intermittently expressed its concern about U.S. troop levels in Europe. Then, it was Senator Mansfield leading the demand for reductions. Now, in the mid-1980s, it is senators Stevens and Nunn who are worrying the administration with their call for phased reductions unless the Europeans meet their NATO defense budget commitments.

In 1978 and 1979, the Senate insisted that President Carter accept a nuclear-powered aircraft carrier he had not requested and did not want. In the 1980s, President Reagan has met persistent resistance from Congress both to his MX missile program and to his Central American policy. Success in achieving the president's target of 100 missiles deployed in strengthened Minuteman silos is by no means assured, and his support for the "contra" guerrillas in Nicaragua was once seriously threatened. In the end, presidents usually manage to get most of what they want; but they have to fight for it.

Speaking generally, political parties in the United States are a less powerful influence on defense policy-making than they are in the United Kingdom. But the president is certainly constrained by the power of Congress—and a good deal of energy is expended on winning friends there. However, because cross-voting is so normal, the president is able to devote somewhat less attention to his party than his prime minister equivalent.

The importance of all this is clear. For years it has been fashionable to interpret decisions and policies in the defense field as being fundamentally rational responses to international situations and strategic problems. Not that the international situation is irrelevant; but internal factors are at least as important because policymakers spend as much time looking behind them as they do in front of them, and any adequate explanation of defense policy must take this into account. Although it has become fashionable, for example, to speak of the "strategic dialogue" between the United States and the Soviet Union and to seek explanations of their respective military policies in terms of reciprocated reactions, the defense policies of both superpowers are to some extent determined from within their respective organizations. Thus, though it is customary to speak of the arms race between the United States and the Soviet Union, it is worth considering the possibility that there is no race—just two military bureaucracies, each doing its own thing. Historically these bureaucratic constraints have been neglected, and it is only comparatively recently that political scientists have begun to appreciate their importance.[41]

The underlying thrust of this is that policymakers are rarely as free as their critics think they are. That does not mean that it is necessary to adopt a determinist philosophy that reduces all human action to that of puppets jerking to the whim of blind, impersonal forces, or actors going through the motions of a Greek tragedy. Perhaps it is the "environmental possibilists" who have most nearly got it right with their view that people do indeed exercise freedom of choice but only within a framework prescribed by their environment—and sometimes the environment is very rigid. The constraints outlined in this chapter all impinge on those who have to make policy, and, under their pressure, freedom of action may shrink to "mitigating the bitterness of the unavoidable."[42]

## NOTES

1. D. Kelly, *The Ruling Few* (London, Hollis and Carter, 1952), 1; and J. F. Kennedy once wrote "there will always be the dark and tangled stretches in the decisionmaking process—mysterious even to those who may be most intimately involved." See his foreword to T. C. Sorensen, *Decisionmaking in the White House* (New York: Columbia University Press, 1964).
2. H. Heclo, "Review Article: Policy Analysis," *British Journal of Political Science*, 2, pt. 1 (Jan. 1972): 84; and A. Etzioni, *The Active Society* (New York: Free Press, 1968), 252.
3. Ibid., 85.
4. S. P. Huntington, *The Common Defense: Strategic Programs in National Security* (New York and London: Columbia University Press, 1961), 3–4.
5. M. Desai, "Social Science Goes to War," *Survival* (March–April 1972): 64.
6. A. Wolfers, *Discord and Collaboration: Essays on International Politics* (Baltimore: Johns Hopkins University Press; 1962), 37.
7. Quoted by J. Eayrs, *Fate and Will in Foreign Policy* (Toronto: CBC Publications, 1967), 29–30.
8. L. Tolstoy, "Second Epilogue: A General Discussion on the Historian's Study of Human Life, and on the Difficulty of Defining the Forces That Move Nations. The Problem of Freewill and Necessity," *War and Peace*, trans. L. and A. Mander (London: Macmillan and Oxford University Press, 1943), 1334.

9. A. N. Sabrosky, "The Defense Policy of France," in D. J. Murray and P. R. Viotti, eds., *The Defense Policies of Nations* (Baltimore and London: Johns Hopkins University Press, 1982), 230.

10. Quoted by J. Frankel, *The Making of Foreign Policy* (London and New York: Oxford University Press, 1963), 151.

11. S. Hoffmann, *The State of War* (New York and London: Praeger, 1965), 162.

12. Sabrosky, "The Defense Policy of France," 232.

13. Quoted by W. R. Heaton, Jr., "The Defense Policy of the People's Republic of China, in Murray and Viotti, eds., *The Defense Policies of Nations*, 423.

14. Sabrosky, "The Defense Policy of France," 241; ibid.; ibid.; and ibid.

15. D. Greenwood, "Economic Constraints and the Defence Effort," *RUSI Journal* 113, no. 652 (Nov. 1968): 329.

16. D. Greenwood, "Economic Constraints and Political Preferences," in J. Baylis, ed., *Alternative Approaches to British Defence Policy* (London: Macmillan, 1983), 32.

17. P. Darby, *British Defence Policy East of Suez 1947–68* (London: Oxford University Press, 1973), 108. I am indebted to my colleague John Baylis for drawing my attention to this example.

18. See J. Howorth and P. Chilton, eds., *Defence and Dissent in Contemporary France* (London and Sydney: Croom Helm, 1984).

19. Sorensen, *Decisionmaking in the White House*, 45.

20. Nicholas Ashford, "At Last Defence Comes Under Attack," *The London Times*, 23 May, 1985: 14.

21. R. Hilsman, *The Politics of Policymaking* in *Defense and Foreign Affairs* (New York: Harper & Row, 1971), 4; G. T. Allison, "Conceptual Models and the Cuban Missile Crisis," in W. F. Hanrieder, ed., *Comparative Foreign Policy* (New York: David McKay, 1971); and ibid., 328.

22. See Charles E. Lindblom, *The Intelligence of Democracy* (New York: Free Press, 1965), 137–43.

23. J. P. Cornford, "The Illusion of Decision," *British Journal of Political Science* 4, pt. 2 (April 1974): 233.

24. Ibid., 362.

25. R. Hilsman, *To Move a Nation: The Politics of Foreign Policy in the Administration of J. F. Kennedy* (New York: Doubleday, 1967), 5.

26. Lindblom, *The Intelligence of Democracy*, 144.

27. Chapman Pincher, *Inside Story* (London: Sidgwick and Jackson, 1978), 306; and Z. Brzezinski and S. P. Huntington, *Political Power, USA/USSR* (New York: Viking Press, 1965), 225.

28. Hilsman, *To Move a Nation*, 5.

29. F. Hopkins, quoted in B. Reed and G. Williams, *Denis Healey and the Policies of Power*, (London: Sidgwick and Jackson, 1971), 199.

30. P. Y. Hammond, "Super Carriers and B-36 Bombers: Appropriations, Strategy and Politics," in H. Stein, ed., *American Civil-Military Decisions: A Book of Case Studies* (Birmingham: University of Alabama Press, 1963), and V. Davis, *The Admirals' Lobby* (Chapel Hill: University of North Carolina Press, 1967); ibid., 228; and ibid., 229.

31. D. Holloway, *The Soviet Union and the Arms Race* (New Haven and London: Yale University Press, 1983), 113.

32. Harriet Fast Scott, "The Soviet Decision-Making Process for National Security Policy," in R. L. Pfaltzgraff, Jr., and Uri Ra'anan, eds., *National Security Policy: The Decision-Making Process* (Harnden, Conn.: Archon Books, 1984), 106; and ibid., 106.

33. H. Harding, "The Evolution of Chinese Military Policy" in F. B. Horton III, A. C. Rogerson, and E. Warner III, eds., *Comparative Defense Policy* (Baltimore and London: Johns Hopkins University Press, 1974) pp. 220–21.

34. Sabrosky, "The Defense Policy of France," 245.

35. G. Williams, F. Gregory, and J. Simpson, *Crisis in Procurement: A Case Study of the TSR-2* (London: Royal United Service Institution, 1969), 61.

36. P. Gretton, *Maritime Strategy: A Study of Defence Problems* (New York: Praeger, 1965), 118; and K. Booth, "Navies and Foreign Policy," unpublished paper (1974).

37. Allison, "Conceptual Models and the Cuban Missile Crisis."

38. S. Huntington, "Strategic Planning and the Political Process," *Foreign Affairs* 38, no. 2 (Jan. 1960): 289. For a thoughtful and sophisticated discussion of "rational" and "nonrational" decisionmaking models, see Y. Dror, *Public Policy Making Re-examined* (San Francisco: Chandler Publishing Co., 1968), particularly pp. 129–62.

39. R. T. McKenzie, *British Political Parties: The Distribution of Power within the Conservative and Labour Parties*, 2nd ed. (London: Heinemann, 1967), 635; and R. Butt, *The Power of Parliament* (London: Constable, 1967), 292.

40. M. R. Gordon, *Conflict and Consensus in Labour's Foreign Policy, 1914–1965* (Stanford, Calif.: Stanford University Press, 1969), 281–2; and A. Verrier, "Defence and Politics after Nassau," *Political Quarterly* (July–Sept. 1963): 272.

41. One attempt to interpret Soviet defense policy in terms of Soviet decision-making processes is M. P. Gallagher and K. F. Spielmann, Jr., *Soviet Decisionmaking for Defense* (New York: Praeger, 1972).

42. S. Hoffmann, *The State of War,* 171.

## 2

# *U.S. Defense Policy*

PHIL WILLIAMS

## *Introduction*

The U.S. military buildup in the first half of the 1980s has been interpreted in very different ways. On the one side, there are those who accept the Reagan administration's contention that the program is simply a response to a decade of neglect in which American military capabilities were allowed to fall into a state of unpreparedness greater than at any time since the late 1940s. In this interpretation, the modernization of America's conventional, and especially naval, forces was postponed by the demands of the Vietnam War. The Reagan administration's goal of a six-hundred-ship navy, therefore, is no more than a long overdue refurbishment of American power, made all the more vital by the dynamism that characterized Soviet procurement through the late 1960s and the 1970s. Similarly, the strategic modernization program is seen as a necessary response to Soviet programs which on certain indicators of balance have put Moscow in a superior position.[1]

On the other side, there are critics who argue that the Reagan administration is attempting to restore the kind of military supremacy over the Soviet Union that the United States enjoyed in the 1950s and 1960s. There have been assertions, for example, that the Pentagon is attempting to regain a first-strike strategic capability against Soviet forces.[2] It has also been suggested that the Reagan administration's conventional buildup is equally ambitious and attempts to go beyond the military missions required by containment.[3] If this is so, then it is a fundamental departure from past American defense policy, for although there have been changes of emphasis from one administration to the next, these have essentially been within the overall framework of the containment policy and the means of achieving it most effectively.

Containment itself was clearly established as the priority goal of American foreign policy in the late 1940s when the Truman administration committed itself to the security of Western Europe through the Marshall Plan, the North Atlantic Treaty, and the Military Assistance Program. The policy was both militarized and extended to Asia as a result of the North Korean attack on South Korea in June 1950. The invasion was interpreted by an administration acutely conscious of the dangers of appeasement as a test of its will and determination. It was also regarded as evidence that threats of strategic retaliation might no longer be sufficient to deter the Soviet Union from aggressive moves against America's allies now that Moscow had broken the U.S. monopoly of atomic weapons. Parallels between divided Korea and divided Germany made these alarmist assessments of the Soviet threat to Europe appear compelling and strengthened the arguments of those who advocated a much more vigorous rearmament effort.

As a result, the Atlantic alliance was transformed from a simple guarantee pact into a semi-integrated military organization with a military headquarters in the form of SHAPE, a supreme allied commander in Europe, the deployment of six divisions of American troops in Europe, and the beginning of German rearmament.[4] Although there had been some tentative moves toward the militarization of the Atlantic alliance before June 1950, it is one of the ironies of the postwar security arrangements in Europe that they were, in large part, a response to what could have been construed as simply a civil war on the Korean peninsula. There was an element of self-fulfilling prophecy in this: the establishment of NATO and the subsequent integration into the alliance of a rearmed Western Germany institutionalized an interpretation of the Soviet threat as primarily military in nature. With the creation of the Warsaw Pact and the buildup of conventional and nuclear forces by both sides, the political division of Europe was given military overtones that all the subsequent proposals for disengagement, nuclear-free zones, and mutual force reductions have been unable to dispel.

The American involvement in Korea, and especially the entry into the war of Communist China, also encouraged the United States to extend a policy that, in Europe, had been enormously successful, to a region where conditions were far less propitious. One of the more immediate consequences was a transformation in the American approach toward the conflict in Indochina: "The original disapproval of France as an imperial power disappeared altogether in a redefinition of the French as defenders of the West," while there was simultaneously a growing antipathy toward the "nationalist rebels" led by Ho Chi Minh, who "were steadily amalgamated with the Communist threat."[5] Indeed, it was during the early 1950s that the seeds of American involvement in Vietnam were sown. Material aid was given to the French in Indochina, and attempts were made to establish a bulwark against communism in Asia with a series of bilateral and multilateral alliances. These included defense agreements with Japan, South Korea, and Taiwan, as well as the Southeast Asia Collective Defense Treaty of 1954. Nor were the United States commitments extended only to Asia. The Middle East became another area of concern, particularly after the Suez crisis of 1956. Fear of Communist incursions in this area led to the formulation of what was known as the Eisenhower Doctrine. This was embodied in a congressional resolution of 1957 stating that the United States was "prepared to use armed forces to assist" any nation in the area "requesting assistance against armed aggression from any country controlled by international communism."[6]

The role of world policeman dedicated to combating communism wherever it appeared culminated in the Vietnam War, which, in turn, provoked a reaction among members of the United States public and Congress against what was increasingly regarded as indiscriminate entanglement in world affairs.

The foreign policy of President Nixon was designed to accommodate the resulting pressure for a more limited international involvement. Indeed, the fundamental assumption of the Nixon Doctrine was that although the United States could not retreat from the world, it could play a less demanding and more discriminate role within it.[7] This was to be made possible by America's allies bearing a greater responsibility for their own security and by attempts to establish a more accommodating and less hostile relationship with both Moscow and Peking.

The United States, of course, did not cease to be concerned about the threat posed by the Soviet Union; nor did containment cease to be the dominant theme of American foreign policy. The means of achieving this objective, however, were very different from those that had obtained in the past. The Nixon-Kissinger détente policy was essentially an attempt to compensate for the loss of American will by encouraging the Soviet Union to engage in a policy of what might be described as self-containment.[8] Although the Nixon Doctrine emphasized strength as well as negotiation, in reality negotiation was deemed essential to persuade the Soviet Union to refrain from exploiting American weakness and loss of will. Kissinger's notion of encouraging Soviet restraint by creating a web of interdependent relationships that would give Moscow a vested interest in stability and restraint was an ingenious attempt to compensate for American weakness.

Insofar as this approach required a judicious mixture of carrots and sticks, however, it was enormously difficult to implement. The Congress made it impossible for Kissinger to provide the incentives for Soviet good behavior when it passed the Jackson-Vanik Amendment linking most favored nation status for the Soviet Union in trade matters to increased Jewish emigration. And it took away the stick when it prohibited covert aid by the United States to the forces of UNITA and the FNLA in the Angolan civil war, thereby minimizing the risks for Soviet and Cuban involvement.

Congress was not the only problem, however. Equally if not more significant was the fact that the Soviet concept of détente differed substantially from that of the United States. For Moscow, superpower détente involved American acceptance of the Soviet Union as an equal with equal rights and opportunities to exert military power in those places where Soviet interests appeared to demand it. Nor did détente in any way imply abandonment of "peaceful coexistence" with its connotations of continued competition in those areas where this would not lead to superpower confrontation. It was almost inevitable, therefore, that the United States would become very disillusioned with détente. Soviet activities in the Third World were regarded by the United States as breaching the rules that had been established in the early 1970s, whereas in Moscow's conception these actions were quite legitimate.[9]

Although this discrepancy between American expectations and Soviet behavior provoked a reaction in the United States, the restoration of the more traditional forms of containment was not immediate. Indeed, the Carter administration's most fundamental failure in foreign policy was its inability to sort out its policy toward the Soviet Union, a failure Stanley Hoffmann has characterized as leaving a large hole in the donut.[10] This was apparent, for example, in the arguments between Secretary of State Cyrus Vance and National Security Adviser Zbigniew Brzezinski over the extent to which strategic arms control should be linked to Soviet good behavior in Africa. By the last year of the Carter administration, however, Brzezinski's hard-line approach was clearly in the ascendancy.[11] Indeed, in the aftermath of the Soviet invasion of Afghanistan, Carter made clear that the Persian Gulf was a region of vital strategic importance to the United States and that Washington was prepared to use force to protect its interests in the area. Containment was not only alive and well but was extended into a region in which the United States had previously been able to rely on surrogates, first Britain and then the shah of Iran.

There are doubts, however, about whether the United States has the capability to uphold these commitments in the event that they are challenged. Closely related is the difficulty of establishing priorities among America's many commitments. Indeed, these are just two of the problems that continuously arise in attempts to translate the deceptively simple notion of containment into operational policies. Such problems have been evident throughout the postwar period and have posed American defense planners and policymakers with a series of dilemmas, many of which have resisted final resolution. In the first place, there have been difficulties in attempting to assess the nature of the threat posed to American security by the Soviet Union. Should most importance be attached to Soviet intentions or to Soviet capabilities? Was the Soviet challenge a political or military one? Was it necessary to match the Soviet Union option for option in order to retain a credible deterrence posture? Could the United States move beyond confrontation to establish a more moderate and regularized relationship with Moscow? And what about other challenges to U.S. interests? Were such challenges essentially indigenous and of no relevance to the superpower competition, or should they be understood and dealt with within a bipolar framework?

There has also been continuing disagreement about how the threat could best be dealt with. The role of nuclear weapons in meeting the Soviet challenge has been a particularly controversial matter. Arguments have concerned the most appropriate strategies to deter nuclear attack on the United States itself as well as the effectiveness of nuclear weapons in deterring non-nuclear aggression. Although these issues are intrinsically important, they have been given added significance by their considerable impact, not only on the proportion of national resources devoted to defense, but also on the allocation of resources and the delineation of tasks within defense. There is no easy answer to the question, "How much is enough?" The way in which successive administrations have attempted to answer this question as well as those about nuclear weapons must now be examined.

## *Commitments and Capabilities*

One of the major tasks of statesmanship is to establish and maintain a harmonious relationship between commitments or political obligations and the means to honor or fulfill them. For Washington, this has necessitated careful and continuing assessments of the utility and limitations of military power in supporting its foreign policy undertakings. Not surprisingly, consensus has sometimes been absent, and the willingness of succeeding administrations to allocate resources to military purposes has varied considerably.

Until the outbreak of the Korean War, the Truman government relied primarily upon economic aid in the hope of restoring the Europeans to a position where their own efforts might suffice to attain their security. This reliance was reflected not only in the Marshall Plan but also in the North Atlantic Treaty, which was intended in part to promote a sense of confidence in Western Europe without which social reconstruction and economic progress would have been seriously undermined. As Robert Osgood has argued, the alliance "was intended to provide political and psychological reinforcement in the continuing political warfare of the Cold War."[12] Thus the treaty was the natural concomitant of Marshall aid and was designed to reassure the Europeans as much as deter the Russians. It was felt that the immediate Soviet threat was more political than military, that internal unrest, subversion, and diplomatic pressure were greater dangers than armed attack. As Osgood notes, "There was no significant fear of a massive Russian invasion."[13] Although the Europeans were expected to rebuild their military strength, therefore, economic recovery was to have first priority.[14]

Given this view of the Soviet threat, military power could have only a limited role in containment. Until the Korean War, President Truman resisted all pressure for expanding the defense budget, and has been criticized for allowing political commitments and military capabilities to become increasingly divergent. Thus, one commentator has claimed that "foreign policy and military policy were moving in opposite directions," with American involvement in the world deepening while the defense budget was declining.[15] The ceiling on the budget was, in fact, reduced from $15 billion to $13 billion prior to the Korean War. Yet this reduction reflected the president's conscious priorities. In the light of his commitment to avoid "deficit spending," an increased allocation of resources to defense would have deprived other government policies, including economic assistance to America's allies, of the funds necessary for their successful implementation.[16] Truman was unwilling to cut those programs on the basis of his assessment of the international situation and how it could best be handled.

Not even a report from some of his senior officials, NSC-68, emphasizing the urgent need for conventional rearmament, could shake the president's resolve.[17] The Korean War, however, stimulated rearmament on a scale that even the most implacable opponents of the Soviet Union had been reluctant to contemplate. Its imprint on American foreign and defense policy was profound. The departure from isolation had been marked by the Truman Doctrine of 1947, but "despite its sweeping language, the Truman Administration between 1947 and 1950 had neither the intention nor the capability of policing the rest of the world . . . the real commitment to contain communism everywhere originated in the events surrounding the Korean War not the crisis in Greece or Turkey."[18] As discussed above, Korea led to both the militarization and the extension of containment. It also led to the first attempt to harmonize commitments and capabilities, a process that was to be achieved through a virtual quadrupling of defense spending.

The more alarmist feelings, however, gradually disappeared, and a reduction in the scale of the American defense effort occurred as it became obvious that the North Korean attack was not the precursor to an all-out Soviet offensive against the West. This reduction, which began under Truman, continued under his successor General Eisenhower, who was elected president in 1952. The new government's estimate of the Soviet threat in fact resembled that prevalent in the White House before the Korean War: the danger was seen as long term and indirect rather than immediate and blatantly military.

Furthermore, Eisenhower was deeply wedded to the principles of "fiscal conservatism" propounded most forcefully by Sen. Robert Taft and other members of the conservative wing of the Republican party.[19] Consequently, the president emphasized the need to maintain fiscal and economic integrity for the "long haul."[20] He feared that the Communists were trying to provoke the United States into spending ever-increasing amounts on armaments. The belief that a "bankrupt America" was "more the Soviet goal than an America defeated in war" lay behind Eisenhower's "great equation": security had to be "reconciled with solvency."[21] Underlying government policy was the recognition that, in view of the other demands on resources and the commitment to balanced budgets, higher defense expenditure would demand higher taxation, which might result in a curtailing of private investment in industry. Eisenhower's faith in the private sector of the economy as the key to the nation's well-being prohibited any such alternative. Indeed, "It was widely felt within the incoming administration that, unless the size and rate of defense expenditures were reduced or, at least, stabilized, a vigorous inflation followed by stifling government controls was likely to occur."[22] This would go a long way to creating a "garrison state" in which cherished democratic values would be sacrified on the altar of "national security." The immediate result of such concern was a massive reduction in the defense effort going well beyond the cut projected by the Truman administration. The Korean War had pushed the budget appropriations for defense up to a peak of almost $57 billion for fiscal year 1952. In fiscal year 1955 they were under $30 billion.[23] Although defense appropriations exhibited a gradual trend upward throughout the rest of the fifties, it was not until the Kennedy years that relatively rapid increases were initiated once more.

It is apparent then that in the late 1940s and 1950s there was no straightforward correspondence between the government's overseas commitments and the development of its military capabilities. Yet this is hardly surprising since neither the Truman nor Eisenhower governments saw the threat as exclusively, or even primarily, military except during the brief period of acute tension precipitated by the Korean War. Furthermore, Washington neither expected nor intended to meet all the commitments alone. Although Eisenhower's secretary of state, John Foster Dulles, is often criticized for what is disparagingly termed "pactomania," the alliances he created benefited the United States to the extent that its allies were to bear an important part of the security burden themselves. In the light of such considerations the claims of defense and security were not overwheming and could be moderated against domestic requirements and the needs of the economy. The resulting discrepancy between commitments and capabilities was not treated with the same equanimity by Eisenhower's critics as it was by the president himself. When he was succeeded by John Kennedy in 1961, therefore, an attempt to attain a greater degree of correspondence between commitments and capabilities was almost inevitable. Kennedy's talk of "new frontiers," however, was slightly misleading: containment remained the basic objective of his foreign policy, as it had been of his predecessor's.

It was not so much that there were new frontiers to defend as existing frontiers being defended more strongly and with a greater sense of urgency by the United States. The new administration's intentions were clearly stated in the president's inaugural address, which emphasized the willingness of the United States to incur heavy sacrifices in meeting the Soviet threat. As he put it, America would be prepared to "pay any price, bear any burden, meet any hardship, support any friend, oppose any foe" in the cause of freedom. Strong words were to be backed up by strong actions requiring dramatic increases in military capability.[24]

Underlying the new posture was a more pessimistic assessment of the Communist threat and a more optimistic assessment of what the United States could do in meeting it than had been made by Eisenhower. Indeed, much of Kennedy's defense policy was an attempt to make up what were regarded as shortcomings in the strategic design of the previous administration. The fifties were seen as years in which the leadership of the Western camp had been allowed to languish and American preparedness undermined.[25] Commitments were now to be fulfilled with an eagerness and enthusiasm that would both provide the example for America's allies in framing their own defense efforts and demonstrate to them that Washington could be relied upon to provide substantial military assistance in the struggle against communism. Consequently, security was accorded a much higher status than in previous years, when, in Kennedy's words, "We tailored our strategy and military requirements to fit our budget—instead of fitting our budget to our military requirements and strategy."[26] Increases in the defense budget over that of Eisenhower were considerable: defense appropriations of just above $39 billion in fiscal year 1960 had climbed to over $47 billion by fiscal year 1964.

They climbed even higher as this activist policy culminated in direct American participation in the Vietnam War. With the increasing unpopularity of the war, however, it became obvious that another reappraisal of American foreign and defense policy was called for. The more radical critics of the war hoped that in its aftermath there would be a significant demilitarization of U.S. policy if not the complete abandonment of containment. The policies of Richard Nixon and Henry Kissinger, while not fulfilling all these demands, recognized them as legitimate grievances. President Johnson's successor acknowledged that Vietnam had become a blatant symbol of United States overinvolvement in the world and accepted the need for a reordering of priorities to give greater attention to urgent domestic problems. The Nixon administration quickly realized that the "American domestic consensus" had been "strained by 25 years of global responsibilities" and that social and economic difficulties at home demanded retrenchment abroad.[27]

The extent of the consequent readjustment should not be exaggerated, significant though it was. The government did not relinquish its obligations but expected the allies themselves to do more, particularly in Asia.

The point was neatly cystallized in the president's foreign policy report to Congress in May 1973, which stated that the "United States will keep all of its treaty commitments but we will adjust the manner of our support for our allies to new conditions and we will base our actions on a realistic assessment of our interests.[28] In essence, this meant that although the United States would still be prepared to furnish military and economic assistance to its Asian allies, it would "look to the nation directly threatened to assume the primary responsibility for providing the manpower for its own defense."[29] There was nothing new about this, though, and it did little more than revert to the position held by Eisenhower and Dulles throughout the 1950s. Once again American involvement in a ground war in Asia was something to be avoided.

The most immediate consequences of this changing attitude were felt in Vietnam itself, where indigenous forces began to take over an increasing share of the war from the Americans. As the Vietnamization program proceeded, American military forces totaling well over half a million men were gradually withdrawn from the area. The American presence elsewhere was also scaled down and included a reduction of twenty thousand troops in South Korea and thirteen thousand in the Philippines. Substantial cuts in the defense budget were thereby made possible. Appropriations that had climbed to $71 billion by fiscal year 1969 were reduced to $66.6 billion by fiscal year 1971. The funds no longer required for defense were directed into civilian welfare programs, thus giving rise to what was termed the "peace dividend." This peace dividend was not sufficient to satisfy congressional liberals who felt that Vietnam had distorted national priorities and that Nixon, by cutting social welfare expenditures, had exacerbated the problem. Yet neither this, nor Nixon's emphasis on strength as the necessary counterpart of negotiation with adversaries, should obscure the fact that the early 1970s saw a significant cutback in the American military effort. This was symbolized by the abandonment of the idea that the United States should be able to fight two and one half wars (i.e., two major wars and one minor contingency) simultaneously and its replacement by the more modest requirement that it should be capable of fighting one and a half wars. The effects of this and the accompanying expenditure cuts were all the greater because they came at the end of a period when resources had been diverted into the Vietnam War at the expense of broader modernization programs.

This was recognized by the Ford administration, which reversed the downward trend in defense expenditure. The need for modest increases was also acknowledged by President Carter. Although the Carter administration was severly criticized as unreliable on defense, it was aware that even the lower level of American commitment bequeathed by the Nixon administration required augmented defense expenditures. To some extent, however, the Carter position was obscured not only because he had campaigned on the need to cut the defense budget but also by the new administration's unwillingess to sanction increases in defense expenditure as great as those outlined in Ford's projection.[30] Indeed, Carter was in the unenviable position of being criticized by liberals for increasing the defense budget beyond the previous year while being attacked by conservatives for increasing it less than Gerald Ford had planned to do. Nor did the administration help its own cause: after working hard to get agreement in NATO on the need for a 3 percent per annum real increase in defense budgets, Carter allowed an internal squabble to develop over the baseline on which this 3 percent was to be calculated.[31] When this occurrence was set alongside specific decisions such as the cancellation of the B-1 bomber and the decision not to go ahead with the neutron bomb, it reinforced the belief of those critics on the Right who suggested that Carter was unreliable on defense.

Other aspects of the Carter defense policy had a similar impact. Furthermore, Carter aroused considerable anxieties about the strength and durability of the American commitment to its allies in Asia. The announcement in 1977 that American ground troops were to be withdrawn from South Korea was explicable in terms of Carter's desire to avoid being dragged into a future conflict on the peninsula simply by virtue of the American presence on the ground. Nevertheless, it encountered considerable opposition from the military and the Congress as well as from Japan and South Korea. It was not entirely surprising therefore that in July 1979 Carter's national security adviser, Brzezinski, announced that the withdrawal had been suspended. The Nixon administration had managed to relinquish commitments in Southeast Asia at a time when concern over the impact of global involvement on the American economy was intense; the Carter administration had found it impossible to relinquish an American commitment in northeast Asia in a period of growing concern over American security and power.

Part of the anxiety was that the Soviet Union was being allowed to make gains in the Third World with impunity. The presence of Cubans in Angola and the Horn of Africa compounded these worries, while the fall of the shah, although it had nothing to do with the Soviet Union, further underlined the need to develop capabilities specifically designed to respond to challenges in the Third World. In fact, the idea of such a force had been discussed in the Carter administration as early as 1977 and was one of the themes dealt with in Presidential Review Memorandum 10.[32] Until 1979, however, there was little impetus behind the plan. But by June 1979, it was clear that the Policy Review Committee of the National Security Council was examining new approaches to Western security in the Persian Gulf, and in the aftermath of this appraisal, Carter decided to set up a joint task force. By December this was known as the Rapid Deployment Force (RDF) and given added importance by the Iranian seizure of the hostages and the Soviet invasion of Afghanistan.

The promulgation of the Carter Doctrine for the Persian Gulf in Janaury 1980 further underlined the importance of the RDF. Yet there were still concerns that the United States had taken on a new commitment without the capabilities necessary to maintain it. Although funds were devoted to increased air- and sealift as part of a substantial increase in the defense budget in Carter's last year, the forces earmarked for the RDF were also earmarked to reinforce Western Europe in the event of a confrontation between NATO and the Warsaw Pact. As a result some critics have argued that any American intervention in the Persian Gulf would undermine the credibility of its commitment to Europe.

A further twist has almost certainly been added to this problem by the Reagan administration's concern with Central America. The Kissinger Commission set up by Reagan to help create a bipartisan consensus on American policy advocated that the nation's commitments there be given the same degree of significance as those in Western Europe, Asia, and the Persian Gulf. The long-term consequence of such emphasis could be to widen the gap once again between America's commitments and its capabilities to meet them. Even the Reagan administration's ambitious spending program—which has increased defense from one-quarter to around one-third of federal expenditure—seems unlikely to be sufficient. Despite the funds allocated to the RDF, for example, crucial weaknesses will remain. Only very few units will be capable of rapid deployment; the rest will take weeks rather than days to arrive and even then may be dependent on the availability of bases and other facilities. The overall implication is that the gap between American commitments and capabilities may be widening rather than narrowing.

Almost certainly this will lead to intensified demands for greater burden-sharing by America's allies, especially in Europe. It will also give the debate over American defense priorities an added intensity. Reagan's first term saw a considerable shift of emphasis away from the dominant themes of the Carter years. Whereas Carter's main concern seemed to be the American commitment to Western Europe, the trend under Reagan has been toward a more globalist approach. The Reagan administration has given priority to maritime capabilities and has emphasized the importance of the Pacific region for American prosperity. This shift toward the Pacific should not be seen as presaging an American withdrawal from Western Europe—America has long been a Pacific as well as an Atlantic power and since 1945 has fought two large-scale wars in Asia without abandoning Western Europe. Indeed, it has been estimated that over 50 percent of the American defense budget is related to the American commitment to Western Europe. Nevertheless, it does seem likely that in the future the United States will increasingly emphasize a "maritime strategy" as opposed to "coalition defense."[33] Such a trend could be fueled by American impatience with European allies who criticize Washington's policies toward Moscow and seem reluctant to devote sufficient resources to their own defense. It may become even sharper as the United States increasingly finds itself confronting acute resource allocation dilemmas. Although there are possible trade-offs between conventional and nuclear weapons that would enable the United States to avoid downgrading its forces in Europe, the choices seem increasingly to favor the nuclear elements. Indeed, the United States has adopted nuclear strategies that, if taken to their logical conclusion, seem likely to demand ever-increasing resources.

## *American Nuclear Strategy*

The idea of nuclear war-fighting as a central feature of United States deterrence strategy can be traced back at least as far as the early 1960s. In 1962, Secretary of Defense Robert McNamara, in a famous address at Ann Arbor, Michigan, suggested that the best way to deter nuclear war was to approach it in much the same way as the wars of the past and make the adversary's military forces the primary targets. At the same time, he argued that in a war the United States would have to keep sufficient forces in reserve capable of hitting Soviet cities to ensure that the Soviet Union too had an incentive to forgo strikes against American population centers. In subsequent years it appeared that McNamara had retreated from this position, as the counterforce strategy, at the declaratory level at least, was superseded first by a broader concern with damage limitation and then by the notion of assured destruction. Indeed, the prevailing assumption in the latter half of the 1960s was that the United States had embraced an assured destruction strategy in which deterrence was based on its capacity to destroy a certain percentage of the Soviet population and industry even after absorbing a first-strike attack. Most commentators, not surprisingly in view of McNamara's pronouncements, failed to make a distinction between assured destruction as a declaratory strategy that established criteria for acquisition policy (thereby giving a handle on planning and helping to answer the question how much is enough) and United States targeting strategy enshrined in the Single Integrated Operational Plan (SIOP), which continued to place considerable emphasis on Soviet military targets.[34] As Desmond Ball has pointed out, the modifications in strategic policy after 1962 "related only to the declaratory and force posture aspects of the policy. McNamara never abandoned his belief in the necessity for flexibility and options, and no substantive changes were made to the 1962 SIOP."[35]

Consequently, the changes initiated by the Nixon administration and publicly enunciated by Secretary of Defense James Schlesinger were less dramatic than was suggested by both officials and observers at the time. The idea that Schlesinger replaced a targeting strategy designed to inflict assured destruction on the Soviet Union with one limited to counterforce attacks is false. Options for both kinds of targets had been included in the SIOP since 1962—and even under Schlesinger the United States was careful to retain a capacity to hit Soviet cities.

This is not to deny that there were significant changes in the early 1970s. One the contrary, several important modifications were made. In particular, Schlesinger emphasized the need for the United States to have a capacity for limited nuclear options and to adopt a more selective and discriminate approach to its targeting. Flexible nuclear response, as it was sometimes called, required a comprehensive range of options sufficient to ensure that, whatever the level of hostilities, the United State could not be coerced: the president should never be put in a position in which his only options were either backing down or suicidal escalation. Consequently, National Security Decision Memorandum 242, signed by Nixon in January 1974, "directed that further plans 'for limited employment options which enable the United States to conduct selected nuclear operations' be developed and formally incorporated into the SIOP."[36]

An additional concern was that the United States should not be worse off than the Soviet Union at the end of a nuclear conflict. As a result, considerable attention was given in subsequent policy guidance documents to a separate category of Soviet economic targets designated "recovery resources." The modified SIOP, which took effect in January 1976, was designed to ensure that the United States could destroy 70 percent of the industrial resources that the Soviet Union would need for its economic recovery after a superpower nuclear exchange. It was hoped, of course, that hostilities would be ended before such strikes became necessary. Indeed, limited nuclear options only made sense on the assumption that intrawar deterrence would prove effective in preventing both sides from initiating large-scale strikes against each other's cities. As Ball has noted, "The concept of 'escalation control' was central to the policy outlined." Equally important was the idea of "withholds," that is, targets such as the centers of political leadership would be exempted from destruction so long as there was a prospect that the belligerents could retain a degree of control over their exchanges.[37]

These ideas underwent further refinement during the Carter administration and were officially approved in Presidential Directive 59 (P.D.59), in July 1980. The approach, which became known as the countervailing strategy, was explained most succinctly by Walter Slocombe (deputy under secretary of defense for policy planning in the Carter administration) in an article in *International Security.* The main feature of the strategy, according to Slocombe, was:

> the proposition that deterrence over the full range of contingencies of concern requires in an age of strategic parity that the United States have forces, and plans for their use, such that the Soviet Union, applying its

> own standards and models, would recognize that no plausible outcome of aggression would represent victory on any plausible definition of victory. In short, this policy dictated that the United States must have countervailing strategic options such that, at a variety of levels of exchange, aggression would either be defeated or would result in unacceptable costs that exceed gains.[38]

The logic was the same as that of Schlesinger and McNamara: the need for an assured response. Credible deterrence of less than all-out attacks required a capacity for less than all-out retaliation. Indeed, only if the United States had an obvious capability for a "selective and measured" riposte would it be able to deter limited attacks. At the same time, the likely response would still be a daunting one for the Soviet Union, as it would "exact a high cost from the things the Soviets value most—political and military leadership and control, military force both nuclear and conventional, and the industrial economic capacity to sustain military operations."[39]

If such a strategy was to be effective, the United States had to not only incorporate greater flexibility into its strategic forces, but also improve their endurance—especially that of command, control, and communications facilities—in case the war dragged on. As well as planning for limited nuclear war, it was necessary to plan for protracted war in order to convince Soviet leaders that even a prolonged conflict would bring them neither victory nor advantage. In accordance with these ideas a new so-called Nuclear Weapons Employment Policy was drawn up. "NUWEP-2 gave priority to targeting Soviet military capabilities, including both nuclear and conventional forces, and to targeting the Soviet military and political leadership, and within the category of economic targets, to war-supporting industry rather than to Soviet economic recovery capabilities."[40]

NUWEP-2 was superseded by NUWEP-82, issued by Reagan's secretary of defense, Caspar Weinberger, in July 1982. Although there does not seem to have been a significant change in targeting priorities, the philosophy and the declaratory position did change considerably. The disclosure of a number of Pentagon planning documents in 1982 suggested four main requirements for American strategic forces:

(1) The United States must have the capacity to "render ineffective the total Soviet (and Soviet allied) military and political power structure through attacks on the political and military leadership and associated control facilities, nuclear and conventional military forces and industry critical to military power."[41]

(2) It is also essential that the United States possesses nuclear forces "that will maintain throughout a protracted conflict and afterward the capability to inflict very high levels of damage against the industrial and economic base of the Soviet Union and her allies so that they have a

strong incentive to seek conflict termination short of an all-out attack on U.S. cities and economic assets."[42]

(3) The United States must "maintain in reserve under all circumstances, nuclear offensive capabilities so that it would never emerge from a war without nuclear weapons while still threatened by enemy nuclear forces."[43]

(4) Command-and-control facilities have to be improved to a point where they are capable not only of "supporting controlled nuclear attacks over a protracted period" but of maintaining links with those submarine-launched ballistic missiles (SLBM) forces that would be held in reserve throughout the conflict.[44]

In other words, the United States would be as concerned with limiting damage to itself as with inflicting destruction as such on the Soviet Union. This is an approach to nuclear war as something akin to conventional warfighting. Furthermore, it is presumed that negotiations would continue throughout the conflict both as part of a coercive bargaining process and as a means of terminating hostilities. The United States must also ensure that the conflict is ended on reasonably satisfactory terms. At a minimum, it must not be left in a position, either in the closing stages of the war or in the postwar world, where it could be coerced by the Soviet Union. In short, even in a protracted nuclear war Washington must be able to prevail.

One problem with this strategy is its cost. The outgoing chairman of the Joint Chiefs of Staff described the approach as a "bottomless pit" that could all too easily distort defense priorities.[45] What makes such criticism even more pertinent is that notions of prevailing in a nuclear war are almost certainly illusory. Indeed, there have been occasions on which President Reagan has admitted that this is so. Nevertheless, the shift in the rhetoric does not seem to have been accompanied by any change in policy. On the contrary, the administration's emphasis on developing defense systems and antisatellite weapons suggests that it is strongly committed to the idea of prevailing.

As well as the policy's financial costs, there are also considerable problems involved in implementation. It is not clear, for example, whether the United States conceives nuclear war as a coercive bargaining process or as a traditional contest of military strength. The difficulty is that these differing conceptions impose conflicting requirements. In a trial of strength Soviet command, control, and communication facilities ($C^3$) would be primary targets as would the political and military leadership.[46] In a coercive bargaining contest, strikes against these targets would be counterproductive. Indeed, it seems doubtful if the United States could initiate decapitation strikes against Soviet political and military control targets without simultaneously destroying any prospects for the control of hostilities. Even if the United States refrained from attacks against Soviet $C^3$ facilities, these could still suffer considerable collateral damage. The vulnerability of American command and control is an equally formidable problem. As Desmond Ball has pointed out, many $C^3$ installations are "inherently large, fixed, and soft."[47] They are therefore vulnerable not only to direct attack but also to the indirect effects of nuclear weapons. The Reagan administration's substantial investment in $C^3$ will help to ease the problem but is unlikely to eliminate it.

The vulnerability of command-and-control centers is only one of many difficulties in any attempt to limit nuclear war. The fact that decision makers would be under great stress, and that technological faults or human errors could have unpredictable but far-reaching consequences, militates against a limited and protracted conflict.[48] Misperception, misunderstanding, and miscalculation could also disrupt all attempts to control hostilities. It is difficult to disagree with Ball in his conclusion that "nuclear weapons are simply too powerful and have too many disparate effects, not all of which are predictable and some of which may currently not even be appreciated, to be used in precise and discriminatory fashion. Any nuclear attacks of military or strategic significance beyond those intended merely to demonstrate political resolve, are likely to produce millions of casualties."[49] Even a limited strategic war is likely to be an unprecedented disaster, and the likelihood is that such a war would rapidly and uncontrollably escalate. Thus, although both superpowers have rejected strategies based on mutual assured destruction, this is still the most probable consequence of a nuclear war. MAD may not be a doctrine, but it is almost certainly a fact of life.

President Reagan's Strategic Defense Intiative, announced in March 1983, is an attempt to change this fact of life and to move from a deterrent posture based on threats of retaliation and mutual vulnerability to one based on the dominance of defensive systems. In launching his initiative, the president was almost certainly influenced by General Graham of the High Frontier group and by the nuclear physicist Edward Teller. Although Graham and Teller differed about the most appropriate means to achieve their objective, both were fervent advocates of a major program of active defense. The Joint Chiefs of Staff also seemed to have some impact when they voiced their concerns to the president about the potential vulnerability of land-based missiles and the political difficulties of getting public and congressional support for new generations of land-based ICBMs in view of the problems encountered with MX. The president himself, however, went far beyond the obvious idea of silo protection to project a vision of a world in which nuclear weapons are made obsolete.[50] His initiative has posed a conceptual and philosophical challenge to the prevailing orthodoxy and has provoked considerable controversy.

There has been much skepticism about the feasibility of defenses that are anywhere near leak-proof. In response to this skepticism, lower-level officials have developed a rationale for the program in terms of denying the Soviet Union its strategic objectives in a war. It is argued that if the Soviet leadership knows that it cannot achieve its objectives even with a first strike against the United States, then it will be deterred from launching such an attack. Deterrence through denial is seen as more effective than deterrence through retaliation. This idea, however, is far less of a departure from the traditional concept of deterrence than it first appears. Insofar as denying the Soviet Union its objectives would require protection of missile silos, then the intermediate option is not inconsistent with the idea of deterrence through retaliation.

Part of the confusion stems from the fact that the SDI, for the first five years at least, is simply a research program, the final outcome of which is uncertain. Nevertheless, the establishment of the SDI Organization under General Abrahamson and the decision to request $26 billion for research over five years has given the program considerable momentum. Although Congress cut back the funding for fiscal years 1985 and 1986, the administration is eager to ensure that by the time there is a new president, the program will have developed such momentum that it will be unstoppable. The research effort has been organized into five main areas: surveillance, acquisition, tracking, and kill assessment; directed-energy weapons technologies; kinetic energy weapons technologies; systems concept and battle management; and survivability, lethality, and key technologies. In this research emphasis is placed on destroying the adversary's ICBMs at the boost phase, in midcourse, and at the terminal phase.[51]

Although the Reagan administration has justified its program in terms of greater Soviet efforts to create defense of its own, it has not stilled the criticisms that have been leveled at the SDI. Perhaps the most salient concerns have been those related to the impact of SDI deployment on strategic stability.[52] Doubts have been expressed in particular about the effect of the SDI on both crisis stability and arms race stability. In response to this some administration spokesmen, such as Paul Nitze, have emphasized that defenses should only be deployed if they meet certain conditions. They must not be vulnerable to a sudden attack and must be cost-effective when considered against possible countermeasures the adversary might take in order to penetrate them. Although budgetary constraints imposed by Congress seem likely to both slow down the program and to underline the importance of Nitze's conditions, the pressures for deployment will be very considerable. The SDI gives meaning to notions of "prevailing" and offers opportunities for bureaucratic aggrandizement and economic benefit. Although it was very much a personal initiative by President Reagan—and one that enabled him to obtain the moral high ground in responding to the American peace movement—the program seems almost certain to generate its own constituencies. By so doing, it will reinforce the arguments of those critics who see the military-industrial complex as one of the major driving forces behind the United States defense policy. Indeed, it is to the organizational and domestic aspects of defense policy that attention must now be given.

## *Domestic Pressures and Defense Policy*

American policy-making for national security is highly fragmented, enormously competitive, and open to the influence of powerful vested interests both within the military establishment and outside. Although the system is highly complex, three dimensions in particular stand out as particularly important: the role of Congress and the executive, inter-service rivalries, and the impact of the military-industrial complex.

Struggle between the executive and legislative branches of government is endemic in American politics. Often described as a government based on the "separation of powers," the American political system is better understood as a government of "separate institutions sharing powers."[53] This is particularly important in relation to defense policy and foreign affairs since the Constitution provided for the president as commander in chief while giving Congress the power to declare war and to allocate the funds necessary for the maintenance of armed forces. Congressional war powers were never used as widely as was intended, with presidents often circumventing Congress, and the postwar era saw them fall into almost complete disuse. President Truman established the pattern by failing to request a congressional declaration of war either before or after he committed U.S. troops to the aid of South Korea. Although there were protests from Republican senators such as Robert Taft that Truman had behaved unconstitutionally, these had little impact. The accretion of power to the president continued throughout the Cold War, facilitated by the concentration of information and expertise in the executive branch.[54] It was accepted—although not without occasional rumblings of disquiet—so long as the president acted within the limits of a national consensus on policy. When the consensus broke down in the face of disillusionment and dissent over the Vietnam War, Congress attempted to reassert its own rights and powers and restrict those of the president. The struggle was a long and difficult one, however, and it was not until 1973 that Congress finally overrode President Nixon's veto and set a sixty-day limit to any presidential commitment of U.S. troops to hostilities abroad.[55]

The Vietnam War also provoked congressional hostility to increases in the defense budget. This was an important departure, for Congress had typically acquiesced in the president's budgetary demands. Toward the end of the 1950s, it had briefly played a more positive function, and as a result of its initiatives extra funds were set aside for the Strategic Air

Command's dispersal and alert facilities and for speeding up the Polaris and Minuteman missile programs. Such activism was the exception rather than the rule, however, and during the 1960s Congress reverted to a more passive review of defense expenditure.[56] It was not until the Tet offensive of 1968 demonstrated the complete bankruptcy of President Johnson's Vietnam policy that Congress became more assertive in its treatment of the executive's budgetary proposals. "For the fiscal years 1960–69 Congress changed the administration's defense budget request by an average of little more than 1 percent. For the fiscal years 1970 to 1973, when dissatisfaction became more pronounced and concern about budget priorities became more serious, the Congress reduced the administration's defense budget request by an average of 4 percent."[57]

By the late 1970s, however, the respective roles of Congress and the president had reverted to those prevalent in the last years of the Eisenhower administration. President Carter, like Eisenhower twenty years earlier, faced considerable congressional pressure to expand the defense budget to a greater extent than he felt was either necessary or desirable. Indeed, the change in attitudes was most striking in the Senate. The attempts of the early 1970s to cut the defense budget were followed by efforts to persuade the Carter administration to spend more on defense. These pressures became particularly insistent as Carter attempted to get SALT II through the Senate—and in late 1979 the president agreed on a 5 percent real increase in the defense budget as the price he had to pay for his arms control treaty. In the event, of course, SALT II ran afoul of the seizure of the American hostages in Iran and the Soviet invasion of Afghanistan. While these events compelled Carter to defer further consideration of the treaty, however, they also strengthened the arguments of those who advocated even larger increases in defense spending. Indeed, the sharp upward trend in military expenditure that characterized Reagan's first term began toward the end of the Carter administration when the president accepted the need for greater increases than he had hitherto been prepared to contemplate.

Furthermore, during the first Reagan administration, Congress gave the president almost everything he wanted for defense. As a result, the defense budget approved by Congress for fiscal year 1985 was around 50 percent higher than that of 1980.[58] The battle over the fiscal year 1986 budget was more intense with the key question being whether there should be a complete freeze on authorizations at the fiscal year 1985 level or an increase sufficient to keep pace with inflation. Whatever happens to authorizations, however, outlays will continue to increase through the remainder of the 1980s because of the backlog of unspent funds. Nevertheless, it is clear that the defense bonanza is coming to an end. Congress has also become much more assertive on particular weapons systems and has imposed limits on the numbers of MX that can be deployed in existing silos. Although few other weapons systems are either as controversial or as politically vulnerable as the MX, the pressures to restrain the defense budget seem likely to intensify during the next few years as concern mounts over the budget deficit. Reagan's policies suffer from the fact that although he is a fiscal conservative in the Eisenhower mold, his desire to exempt the defense budget from the cutbacks in federal spending contributes significantly to the overall deficit. Even if the administration were much more concerned with pruning defense expenditure, however, it would face immense problems.

The pressures from the services themselves for higher defense budgets are not inconsiderable. Furthermore, if the administration is reluctant to acquiesce in specific military demands, then appeals can always be made to Congress. Throughout the postwar period in fact, Congress has acted as a forum and court of appeal for the armed services in their struggles over the allocation of missions and resources. Although there has been general agreement within the defense establishment on the objectives of United States policy, the role and importance of each of the military services in attaining those objectives have been much more uncertain. As a result "strategic programs are not the product of expert planners who rationally determine the means necessary to achieve desired goals. They are the result of controversy, negotiation and bargaining among different officials and groups with different interests and perspectives."[59] Not even the National Security Act of 1947, creating the post of secretary of defense and supposedly unifying the military establishment, could alter this and significantly moderate the intensity of interservice rivalry.

In the 1940s this rivalry took the form of a dispute over the control of atomic weapons. The newly independent air force sought to maintain a position of exclusive ownership and attempted to establish its preeminence over the older services with the argument that the next war would be won by strategic bombing alone. The fact that from 1947 both the army and navy had their appropriations cut in "each fiscal year until the North Korean attack in 1950" is testimony to the efficacy of its tactics.[60] The two senior services quickly realized, however, that nuclear missions were a means to organizational growth and a greater share of the defense budget, and the navy established a nuclear niche for itself with the strike carriers, and later the Polaris submarines, while the army did the same with the development of tactical nuclear weapons. Despite this, they were unable to dislodge the air force from its well-entrenched position at the top of the military hierarchy.

Budgetary battles continued throughout the 1950s and were particularly serious because of the limited resources made available for military purposes. The effect on procurement decisions and research and development was particularly important as "each service launched its own projects independently of the others in the hope of laying claim to some future mission and thereby increasing its share of available funds."[61] Not surprisingly, there was considerable duplication of functions as a result, with both the army and navy developing their own intermediate-range ballistic missiles. It was when the army lost this particular battle that it became interested in ballistic missile defense (BMD). Close links were established with certain defense contractors, and with the enthusiastic support of key figures in Congress, the army was able to exert sufficient pressure that Secretary of Defense McNamara, who was hostile to the principle of ballistic missile defense, was compelled in 1967 to announce the decision to deploy a thin ABM system.

Although he lost this particular battle, McNamara saw his task in terms of preventing wasteful expenditure. Regarding decentralized decision making as an abdication of the authority of the secretary as well as a source of waste and inefficiency, McNamara tried to make the allocation of resources within defense a more rational process than it had been under his predecessors. Rigorous techniques for evaluating weapons and force levels were introduced and a planning-programming-budgeting system, with its five-year projections of the defense posture, was established as an aid to centralized decision making and long-term planning. One consequence of these innovations was that military judgments about the feasibility and desirability of particular weapons systems were sometimes overruled by the secretary on cost-effectiveness considerations. This earned McNamara considerable criticism from Congress, and under his successors, the service chiefs and the Joint Chiefs of Staff appear to have regained much of their former importance.

In Jimmy Carter's first year, however, there were once again complaints that military advice was being disregarded. It was also alleged that Harold Brown, Carter's secretary of defense, conceived of his role as an agent of the White House rather than as the champion of the services. When budget procedures were reformed to allow the president to participate in the process at an early stage, this further intensified the criticisms. Although the military concerns eased as Brown gradually became a vigorous advocate of higher defense spending, the navy in particular remained unhappy. The administration's preoccupation with strengthening ground forces for central Europe, the decision to cut back on the ship-building program projected by the Ford administration, and Carter's decision in 1978 not to accept the fifth nuclear-powered aircraft carrier alienated the navy even though it received a higher share of the defense budget than either the army or the air force. Under the Reagan administration the navy's slice of the budget has, in fact, declined, and the modernization of strategic programs has put the air force in the top position. Nevertheless, the administration's commitment to the development of a six-hundred-ship navy means that the procurement program will expand substantially throughout the rest of the 1980s and into the

1990s. Indeed, during 1983 Deputy Secretary of Defense Paul Thayer (who later resigned to fight allegations of improprieties in financial dealings that took place prior to his appointment as the second-ranking Defense Department official) argued that the navy was being given too high a priority and that funds should be redistributed to the army. As a result, Secretary of the Navy John Lehman became very active on Capitol Hill in an attempt to forestall any backtracking from the target of a six-hundred-ship navy. Consequently, although tight defense budgets give interservice rivalries a particular sharpness, these rivalries continue even in periods when resources for defense are more plentiful.

In operational terms, these rivalries can prove debilitating. The attempt to rescue the American hostages in Iran is a prime example of the difficulties that can arise during joint operations. Yet there was probably little alternative to this joint involvement given the desire of each of the services to obtain a share of the glory in the event that the operation was a success and to disperse the blame if it was not.

Although these parochial considerations often lead to wasteful duplication of effort, the services are not always enthusiastic about new technologies. During the 1970s, for example, there was considerable reluctance to develop cruise missiles.

> Without exception, the military services did not want cruise missiles if they threatened their respective dominant missions or ate into their scarce funds, both of which were in general the case. The long-range air-launched cruise missile (ALCM) was rammed down the throat of the Air Force. The Army refused to accept development responsibility for the ground-launched cruise missile (GLCM). The Air Force consequently got stuck with it and viewed it as a national mission—one it performed for the country, but that had little value for its own missions. The Navy—specifically the carrier admirals—did not want the Tomahawk antiship missile (TASM) because it represented a clear and present danger to the mission of the carrier-based aircraft.[62]

In contrast to this, any capabilities that promise to strengthen or perpetuate primary roles and missions are fought for with considerable tenacity. Perhaps the best example of this is the B-1 bomber, which has become something of a cause célèbre for the Strategic Air Command (SAC). Although Carter canceled the B-1—largely on the grounds that both air-launched cruise missiles and the new Stealth technologies made it unnecessary, the fact that SAC defines much of its *raison d'être* in relation to the manner bomber helps to explain why it has been revived by Reagan.[63]

The B-1 is also one of the most obvious manifestations of the importance of the military-industrial complex. Although the B-1 has been enormously controversial since its conception in the early 1960s, it refuses to lie down. Its survival has appropriately been described as "testament to the strength of its constituencies in the military-industrial complex more than to any national consensus on the bomber's strategic worth or cost effectiveness."[64] These constituencies include not only the air force itself but also the Rockwell International Corporation, which is building the plane, a myriad of subcontractors including Boeing, Westinghouse, General Electric, Martin Marietta, and Northrop, and those congressmen and senators whose districts and states will benefit economically from its production.[65] The public justification, of course, cannot be in terms of jobs, profits, or votes. Instead, considerable emphasis is placed on the strategic rationales, which, if not completely spurious, are far from compelling. Indeed, the B-1 is one of many instances that reveal the power of those groups with vested interests in exaggerating the Soviet threat. Missile gaps and bomber gaps are made for filling, while windows of vulnerability and opportunity have to be shut. This is not to argue that the Soviet threat is a figment of the defense planner's imagination; it is to suggest, though, that the problem of assessing the threat and what needs to be done to respond to it is one of the major issues facing the United States throughout the latter half of the 1980s and into the 1990s.

## *Current and Future Issues*

The problems facing the United States as it formulates its defense policy in the latter half of the 1980s and into the 1990s can be divided into three major categories: concepts, capabilities and commitments, and consensus.

The conceptual problems revolve around the nature of security and the threat posed by the Soviet Union. In the early 1970s the United States appeared on the verge of embracing a broadened notion of security as something that could best be achieved through unilateral military measures and bilateral agreements with the Soviet Union. Cooperative arrangements based on mutual advantage were regarded as an opportunity to mitigate some of the worst features of superpower rivalry and to reduce the risk of nuclear war. At the very least, it was hoped that a degree of unpredictability could be built in to the continuing competition with Moscow. This approach was undermined, however, by continued improvements in Soviet strategic capabilities, Soviet activities in the Third World, and a general move to the right in American domestic politics. Détente was replaced by concerns over the "window of vulnerability." Furthermore, the anxieties over the emerging vulnerability of U.S. land-based missiles were compounded by fears that such vulnerability would be exploited by the Soviet Union in ways inimical to United States interests. Conservative critics of détente argued that Moscow, aware of its strategic advantage, would be able to take higher risks and to adopt an even more expansionist policy than it already had. In short, the "window of vulnerability" would be accompanied by a "window of opportunity."

The problem with such a view is that it rests upon a "reductionist assessment of the Soviet threat."[66] It "involves an assumption of technological determinism, in the sense that it postulates that the broad aggressive purposes of the Soviets are made operative and given specific content and direction by the Soviet acquisition of new military technologies. Means, in effect, determine goals. By the same token, Soviet acquisition of these technologies is perceived as confirming that Soviet intentions are aggressive and adventuresome."[67] Although the Reagan administration, through the Scowcroft Commission, has tacitly rejected the "window of vulnerability" thesis—at least as an immediate concern—its overall assessment of the threat is still founded on an "inherent bad faith model" of the Soviet Union. This position implies that only unilateral military measures by the United States can contain this threat.

In other words, security is once again defined in narrow and very traditional terms. The difficulty with such an approach is that it heightens tensions and adds fuel to the arms race. Because the Reagan military buildup intensifies Soviet paranoia and thereby provokes a Soviet response, it is unlikely to lead to any lasting gain in security for the United States. This is not to argue that the United States can afford to neglect programs for military modernization. It is to suggest, though, that the broader approach to security that began to emerge in the early 1970s needs to be reinstated. Recognition of the need for constructive engagement with the Soviet Union to enhance strategic stability and reduce the likelihood of superpower confrontation in regions of instability is not inconsistent with American preparedness but is a vital supplement to it. Although negotiated arms control—with its emphasis on parity and the need for bargaining chips—raises more problems than it solves, an attempt to obtain agreement on substantive, as opposed to token, confidence-building measures could have positive benefits for American national security and for international security.

The incentives for moving in this direction are reinforced by the difficulties of reconciling commitments and capabilities. Despite the Reagan administration's military program, it is not certain that the United States will be able to meet all its commitments, especially if there are simultaneous challenges in different regions. Although this can be alleviated by greater burden sharing, it also demands that the military instrument be integrated into a much broader approach. As Robert Osgood has argued, "Some of the most effective things we can do to contain the Soviet threat and bring power into balance with interests lie in strengthening the international environment against Soviet pressure by mitigating the vulnerability of trouble spots, eliciting the cooperation of allies, engaging regional states with convergent interests in diplomatic and economic relationships that restrain local sources of trouble, and collectively addressing the sources of financial and monetary disorder—all with a relatively low American profile."[68] Recognition of the limits of American military power and particularly its lack of suitability in response to a wide range of contingencies is essential if American resources are not to be dissipated in unnecessary and often futile endeavors. In an age of limits the United States can no longer afford to be profligate with resources.

The need for such restraint is reinforced by the fact that American defense expenditures have not always been distributed in the most efficient and effective way. Indeed, during the latter half of the 1970s a military reform movement began to develop in Washington, prompted in large part by the concern that an obsession with technology was distorting America's defense effort.[69] In the view of the reformers—a disparate group of congressional staffers, academics, and military commentators—not only has sophistication been sought at the expense of numbers, but technology itself has become a substitute for imaginative tactics and doctrines. Yet technological innovation is not invariably something to be decried: although sophistication for its own sake has little to recommend it, new technologies may in some instances offer opportunities to implement existing roles and missions more effectively and at lower cost. They may also suggest the need for a reappraisal of existing doctrines and organizational structures. Indeed, if the reform movement does no more than make officials think more seriously about existing rationales, it will have made an important contribution.

The need for greater efficiency in allocating and using resources has loomed more important as the consensus in favor of higher defense spending has been eroded. Several factors have contributed to this erosion. As well as the concern with reducing the budget deficit, there has been a widespread and uniformly hostile reaction to disclosures about contractors charging excessive prices for components. The amount of waste and duplication in the Pentagon and the lack of congruence between the vast resources devoted to defense by the Reagan administration and the modest increases in combat capability have called into question the need for continuing increases in the defense budget. Although authorizations that have already been made will ensure that defense outlays continue to increase for several years, the battle over the defense budget in fiscal year 1986 has shown that President Reagan can no longer expect to get his way on defense spending.

Perhaps even more significant are the question marks about the ability of the American president to mobilize consent for large-scale or long-term military operations. In the 1980s there are two competing impulses in the formulation of American security policy, one of which goes back to Vietnam and the other of which finds its rationale in the ill-fated attempt by President Carter to rescue the American hostages in Iran. The Vietnam syndrome—which still appears to be prevalent in Congress—suggests the need for caution in the use of American military power, particularly in circumstances in which political conditions are not conducive to a quick and decisive outcome. The Iran syndrome, on the other hand, founded as it is on the desire to wipe out the ignominy of the late 1970s, may lead to an excess of assertiveness. How these competing impulses work themselves out will have a profound impact on the future of American defense policy. In the final analysis, that policy can be no better than the underlying assumptions about the world and America's place in it. And so long as the orthodoxy of the new cold war continues to hold sway there are strong grounds for being concerned about the validity of these assumptions.

## NOTES

1. See, for example, C. Coker, *U.S. Military Power in the 1980s* (London: Macmillan, 1983).
2. R. C. Aldridge, *First Strike: The Pentagon's Strategy for Nuclear War,* (London: Pluto Press, 1983).
3. B. Posen and S. V. Evera, "Defense Policy and the Reagan Administration: Departure from Containment," International Security 8, no. 1 (Summer 1983): 3–45.
4. See R. E. Osgood, *NATO: The Entangling Alliance* (Chicago: Chicago University Press, 1962), for a fuller analysis of these changes.
5. J. Stoessinger, *Nations in Darkness* (New York: Random House, 1971), 66.
6. The full text of the resolution can be found in *Global Defense: U.S. Military Commitments Abroad* (Washington: Congressional Quarterly Inc., 1969), 23.
7. See, for example, *U.S. Foreign Policy for the 1970s: Shaping a Durable Peace,* Report to the Congress by President Nixon, 3 May 1973.
8. See S. Hoffmann, *Dead Ends: American Foreign Policy in the New Cold War* (Cambridge, Mass.: Ballinger, 1983), 90.
9. See A. L. George, *Managing U.S.-Soviet Rivalry: Problems of Crisis Prevention* (Boulder, Colo.: Westview Press, 1983).
10. S. Hoffmann, *Dead Ends* 97.
11. The changes in the Carter administration's defense policy are discussed more fully in the author's "Carter's Defense Policy," in G. Abernathy, D. M. Hill, and P. Williams, eds., *The Carter Years* (London: Pinter, 1984).
12. Osgood, *NATO,* 30.
13. Ibid.
14. This is discussed more fully in P. Williams, *The Senate and U.S. Troops in Europe,* (London: Macmillan, 1985).
15. S. E. Ambrose, *Rise to Globalism: American Foreign Policy 1933–1970* (London: Penguin Books 1971), and see also S. P. Huntington, *The Common Defense: Strategic Programs in National Politics* (New York: Columbia University Press, 1961), 20.
16. See P. Y. Hammond, "NSC-68: Prologue to Rearmament," in W. R. Schilling, P. Y. Hammond, and G. H. Snyder, *Strategy, Politics and Defense Budgets* (New York: Columbia University Press, 1962), 278. This volume is an indispensable source of information on the evolution of American defense policy during the later 1940s and early 1950s.
17. See ibid. for illuminating analysis of NSC-68. Also useful are the pieces by J. L. Gaddis and P. Nitze, "NSC-68 and the Soviet Threat Reconsidered," *International Security* 4, no. 4 (Spring 1980).

18. J. L. Gaddis, "Was the Truman Doctrine a Real Turning Point?" *Foreign Affairs* 52, no. 2 (Jan. 1974): 386.

19. Taft's views are discussed in P. Williams, "Isolationism or Discerning Internationalism: Robert Taft, Mike Mansfield and American Forces in Europe," *Review of International Studies* 8, no. 1 (Jan. 1982): 27–38.

20. See W. Schilling, et al., *Strategy, Politics and Defense Budgets*, especially 470–479.

21. Ibid.

22. E. A. Kolodziej, *The Uncommon Defense and Congress* (Columbus: Ohio State University Press, 1966), 108. Such fears are also discussed throughout Schilling, et al.

23. These figures are taken from *Editorial Research Reports on America's Changing World Role* (Washington: Congressional Quarterly, 1974), 15. The figures are based on information supplied by the House Appropriations Committee and the Defense Department and do not include supplementary appropriations for any year.

24. See R. Aron, *The Imperial Republic: The United States and the World 1945–1973* (Englewood Cliffs, N.J.: Prentice-Hall, 1974), 70–79.

25. J. F. Kennedy, quoted in H. Moulton, *From Superiority to Parity* (London: Greenwood Press, 1973), 37.

26. Quoted in ibid.

27. *U.S. Foreign Policy for the 1970s: Building for Peace,* A Report to the Congress by Richard Nixon, President of the United States, 25 February 1971, 7.

28. See *U.S. Foreign Policy for the 1970s: Shaping a Durable Peace,* May 3, 1973, 40.

29. Ibid.

30. See L. Korb, "The Policy Impacts of the Carter Defense Program," in S. Sarkesian, ed., *Defense Policy and the Presidency: Carter's First Years* (Boulder, Colo.: Westview Press, 1979).

31. For a fuller discussion see Foreign Policy Research Institute, *The Three Per Cent Solution and the Future of NATO* (Philadelphia: Foreign Policy Research Institute, 1981).

32. For a discussion of PRM-10 see L. Korb, "National Security Organization and Process in the Carter Administration," in Sarkesian, *Defense Policy.*

33. See R. Komer, *Maritime Strategy or Coalition Defense?* (Cambridge, Mass.: Abt Books, 1984).

34. For the distinctions between different aspects of defense policy see D. Ball, "U.S. Strategic Forces: How Would They Be Used?" *International Security* 7, no. 3 (Winter 1982–83): especially 32–33.

35. D. Ball, *Targetting for Strategic Deterrence, Adelphi Paper* no. 180 (London: International Institute for Strategic Studies, Summer 1983), 14.

36. Ibid., p 19.

37. Ibid.

38. W. Slocombe, "The Countervailing Strategy," *International Security* 5, no. 4 (Spring 1981): 21.

39. Ibid., 23.

40. Ball, *Targetting for Strategic Deterrence,* 23.

41. Quoted in C. Paine, "Nuclear Combat: The Five Year Defense Plan," *Bulletin of the Atomic Scientists* 38, no. 9 (Nov. 1982): 6.

42. Quoted in J. R. Halloran, "Weinberger Angered by Reports on War Strategy," in *New York Times,* 30 May 1982.

43. Ibid.

44. R. Halloran, "Pentagon Drawing Up First Strategy for Fighting a Long Nuclear War," *New York Times,* 30 May 1982.

45. See G. C. Wilson, "Preparing for Long Nuclear War Is Waste of Funds, Gen. Jones Says," *Washington Post.*

46. Paine, "Nuclear Combat," 6.

47. D. Ball, *Can Nuclear War Be Controlled? Adelphi Paper* no. 169 (London: International Institute for Strategic Studies, Autumn 1981), 9.

48. Ibid., 12–14.

49. Ibid., 30.

50. See "Text of Address by the President to the Nation, March 23, 1983," in P. E. Gallis, M. M. Lowenthal, and M. S. Smith, *The Strategic Defense Initiative and United States Alliance Strategy* (Washington Congressional Research Service, 1 February, 1985), 74.

51. SDIO, *Report to the Congress on the Strategic Defense Initiative 1985* (Washington: Department of Defense, 1985).

52. For a good analysis of the problems the SDI might pose, see H. Brown, "The Strategic Defense Initiative: Defensive Systems and the Strategic Debate," *Survival* 27, no. 2 (March–April 1985): 55–64.

53. See R. Neustadt, *Presidental Power* (New York: Wiley, 1960).

54. See J. A. Robinson, *Congress and Foreign Policy Making* (Homewood, Ill.: Dorsey Press, 1967).

55. A useful account of the struggle is contained in J. W. Spanier and E. M. Uslaner, *How American Foreign Policy is Made* (New York: Praeger, 1974), 68–78, 132–43.

56. On congressional influence see Kolodziej, *The Uncommon Defense*. This is usefully supplemented and updated by A. Kanter, "Congress and the Defense Budget: 1960–1970," in *American Political Science Review* 66, no. 1 (March 1972), and by the annual *Congressional Quarterly Almanac*.

57. E. R. Fried, A. M. Alvin, C. L. Schultze, and N. H. Teeters, *Setting National Priorities: The 1974 Budget* (Washington: The Brookings Institution, 1973), 407.

58. See R. Stubbings, "The Defense Program: Buildup or Binge?" *Foreign Affairs* 63, no. 4 (Spring 1985): 848–72, especially 848.

59. S. P. Huntington, *The Common Defense*, 146.

60. See Kolodziej, *The Uncommon Defense*, 65. See also W. R. Schilling's superb analysis, "The Politics of National Defense: Fiscal 1950," in Schilling, et al., *Strategy, Politics and Defense Budgets*.

61. W. Kaufmann, *The McNamara Strategy* (New York: Harper & Row, 1964), 31.

62. R. J. Art and S. E. Ockenden, "The Domestic Politics of Cruise Missile Development 1970–1980," in R. K. Betts, ed., *Cruise Missiles: Technology, Strategy, Politics* (Washington: Brookings Institution, 1981), 360.

63. For Carter's position see J. Carter, *Keeping Faith: Memoirs of a President* (New York: Bantam, 1982), 80–83.

64. D. Wood, "B-1 Symbolizes Power of Military-Industrial Complex," *Los Angeles Times*, 10 July 1983.

65. Ibid.

66. R. H. Johnson, "Periods of Peril: The Window of Vulnerability and Other Myths," *Foreign Affairs* 61, no. 4 (Spring 1983): 951.

67. Ibid.

68. R. E. Osgood, "American Grand Strategy: Patterns, Problems and Prescriptions," *Naval War College Review* (Sept.–Oct. 1983): 16.

69. For a selection of useful analyses of the "reform movement" see "Forum: Military Reform and Defense Planning," *Orbis* 27, no. 2 (Summer 1983): 245–300.

—3—

# *Soviet Defense Policy*

KEN BOOTH

Soviet defense policy must be understood as the product of a unique historical and geographical environment and of a distinct political culture. Together, these factors have helped to create a Soviet-Russian state whose leaders have been particularly security-conscious, whose society has been highly militarized, and whose military outlook has been predisposed toward overinsurance. Environment and experience have produced an ingrained sense of national insecurity, which should be the starting point for any study of Soviet defense policy.

Out of the Soviet-Russian milieu there has emerged a distinctive approach to strategic matters. It is an approach that, despite the pervasiveness of ideology in Soviet life, is more felicitously described as "traditional" than as "revolutionary." This is evident in Soviet thinking about deterrence and arms control, in the Soviet belief that stability is better assured by preponderance rather than equilibrium, and that it is not prudent to rely for one's security on the good intentions of foreigners. Indeed, in contemporary military affairs the true strategic "revolutionaries" and idealogues have not been Soviet marshals. Rather, they have been those nuclear theorists and peace researchers in North America and elsewhere in the West who have expressed faith in the ability of nations to control international conflict by theories based on strategic logic rather than strategic experience, and by the mutual management of armaments policies with sworn enemies. Soviet strategic man has favored more time-worn approaches.

For most of the period since 1917, Soviet strategists have seen their state encircled by technologically more advanced powers; and even though the Soviet Union emerged from the Second World War as the world's second military power, its leaders were painfully aware of American strategic superiority. They were also deeply suspicious of American motives. Consequently, much of what the Soviet Union has done since 1945 can be interpreted as an effort to catch up with its more powerful and dynamic superpower rival, and to discover effective counters. After the most strenuous efforts and at considerable cost, the Soviet Union can now match the United States in military terms; and in some though not all areas, it is ahead. As a result, Soviet leaders now feel more confident in a military sense than ever before. Soviet defense planners are basically satisfied with their country's position. They have reason to believe that the Soviet Union is presently so strong that nobody but a madman dare attack it, and that it is nowhere so weak that it can be ignored in international affairs. In addition, Soviet forces now have global military reach, albeit in limited quantities in some respects; this opens up new possibilities for using the military instrument to advance foreign policy.

Although Soviet defense planners are basically satisfied with their present position, they have many immediate and long-term causes for concern. There is the prospect that the United States is trying to translate its superior economic power into strategic superiority as a result of a costly arms race. There are the problems caused by the modernization of NATO's conventional and nuclear forces. Soviet planners appreciate the complications in strategy resulting from the prospective development of both Chinese and Japanese forces. There is the perennial and probably growing problem of maintaining control in Eastern Europe; and there is recognition of the limitations on the Soviet ability to promote favorable developments in the Third World. All these worries exist against a background of declining manpower for the forces, chronic economic difficulties, and the likelihood of new pressures on the defense budget. The details are novel, but the experience is familiar for Soviet defense planners: they know from the past that new headaches will invariably destroy whatever moments of satisfaction they might have tirelessly earned.

## *The Problem*

From a Western viewpoint, the Soviet armed forces have been the looming ogre in the cold war drama. The military policy that has directed their powerful muscles has often seemed a mystery, though a few analysts have shown that it is possible to discover a great deal about Soviet military policy.[1] Nearly seventy years of word and deed provide plenty of material upon which to work and speculate. This obviously should not be taken to mean that analyzing Soviet military policy is straightforward. It is not. The search for understanding is obstructed by morbid Soviet secrecy, while the inevitable problems of strategic analysis are compounded by a number of attitudes and approaches of Western observers that distort what knowledge is gained. The main difficulties have arisen from a level-of-analysis problem, ethnocentrism, and simple incuriosity.

The level-of-analysis problem expresses itself in confused threat assessment. When we are talking about the "Soviet threat," is the concern that of the military contingency planner with his theoretical range of possibilities of what an adversary can do? Or is it that of the foreign policy analyst, with his estimates of probabilities about adversary behavior? Obviously this distinction between what is possible and what is probable is a crucial one when analyzing the difficult concept of "threat," but it is often ignored or confused.[2] As a result, the simple worst-case forecast of the contingency planner too easily dominates the debate.

Ethnocentrism is the tendency to assess the words and deeds of other groups in terms of one's own culture, and invariably with an implicit or explicit belief in its superiority.[3] In military and other fields the standards and assumptions of one's own outlook are often applied to others in a biased way. Thus one hears that "the Soviets do not understand sea power," when it should be quite clear that they do understand it, but from a different perspective from that of the traditional monopolist maritime nations. Similarly, aspects of Soviet military behavior are sometimes characterized as "not making sense" (such as the institution of the "military commissars"). In such cases it is incumbent on Western strategists to realize that Soviet planners are simply and rationally acting from different assumptions or applying different values.

Cultural blind spots on the part of Western strategists impinge in other ways. For example, if Western analysts do not try to perceive the world from the standpoint of Soviet decision makers, it means that Soviet behavior will be assessed in terms of Western vulnerabilities and fears rather than in terms of Soviet problems and priorities. This will limit our understanding of the "reality" of our strategic predicament. Cultural blind spots also result in a tendency to assume that Soviet spokesmen mean the same thing as their Western counterparts when they use what is translated as the same or similar terminology. This is not so, and not to recognize the fact has led many people over the years to misunderstand the Soviet conception of such an important idea as deterrence.

Understanding Soviet military policy will always be fraught with difficulties.[4] There are both inherent and Soviet-made obstacles, but at least some of the keys are in our own hands. Nevertheless, the fact remains that all but a few members of the Western strategic community have been remarkably incurious about the Soviet "enemy." One of the unfortunate consequences of this has been that few of those responsible for developing mainstream Western strategic studies since the mid-1950s have allowed their outlooks and assumptions to be fertilized by a sophisticated understanding of the adversary's strategic culture.

When it comes to thinking about Soviet defense policy, our ideas about Soviet behavior are still at the mercy of emotions and powerful ideological predispositions, even though we are seven decades from the Bolshevik revolution. This problem has been particularly evident in the Sovietology of the Reagan administration, whose images of the Soviet Union are based on Cold War stereotypes. One effect of the emotional dimension of the subject has been the habit of applying simple labels to Soviet behavior. We are often asked to choose: is the Soviet Union aggressive or defensive? Is it motivated by insecurity or ambition? Or is it driven by ideology or national interest? Such simple labels and dichotomies should be resisted. In detail Soviet foreign policy and strategy have been a tangled, ambiguous, and changing combination of insecurity and ambition, Communist outlooks and Russian national interests, bureaucratic inertia and revisionist opportunism, pragmatic actions and doctrinal rhetoric, defensive-mindedness and revolutionary aspirations, tactical expediency and long-term strategic endeavors, and coexistence and expansion. A great nation's external behavior cannot be reduced to simple labels and dichotomies, as the rest of this chapter will show.

## *The Defensive-Expansionist Heritage*

The invaders of Russia in the last eight hundred years read like a *Who's Who* of military aggression: the Mongols, between 1240 and 1380; the Poles, 1607–12; the Swedes, 1611–14, and again in 1709; Napoleon in 1812; Germany and Austria in 1914–18; Britain, the United States, Japan, France, and Italy, 1918–19; and Germany between 1941–44. The Russian people have faced the scourge of war probably more than any other nation, certainly in Europe.* Within this century the Russians lost forty-five thousand military dead in the Russo-Japanese War; nearly two million military and an uncounted number of civilian casualties in World War I; thirty thousand casualties in the clashes with Japan in the late 1930s; approximately fifty thousand casualties in the Soviet-Finnish War; and, in the last and biggest of Russia's major wars, the country lost about twenty million dead in overcoming the gratuitously brutal invasion by Hitler's Germany.* Outside Leningrad there are no single graves for individual, if unknown, defenders. Each grave contains numerous bodies. The scale of Russian suffering has been heroic; it is almost ungraspable for the traditionally secure and phlegmatic citizens of the British Isles or North America. No wonder, in the Soviet mind, the "Great Patriotic War" was "only yesterday."[5]

Vulnerability to military invasion and pressure has been a characteristic feature of Russian history, and Soviet leaders, like their tsarist predecessors, have never been able to be really confident about their state's security from external threat. The chronic feeling of insecurity in Russian life can be called paranoia, but it has been no less real for being in the mind. In such circumstances, the accretion of military power to deter the state's numerous perceived external enemies, to deal with

*Self-inflicted terror, of course, has also been on a staggering level.

*Estimates put military casualties between 7,500,000 and 13,600,000 and civilian casualties at 7,500,000. The British figures were 264,443 and 92,637 respectively. The United States had slightly more military casualties, and many fewer civilian casualties than Britain. It is significant that Russian war dead are often counted to the nearest thousand, while Britain and the United States are able to do it to the last individual.

them from a position of strength, and to fight and defeat them if deterrence failed[6] had been a constant concern and focus of national efforts. It is not surprising that those responsible for that patch of embattled territory have equated their security with military strength, have defined their security requirements expansively, and have always tended to overinsure in quantitative military terms.[7] Objective and subjective requirements have dictated massive military manpower and equipment.

As well as calling for exhausting military efforts, the Russian-Soviet predicament has also produced a history of expansionism. In tsarist times there was a periodic search for secure frontiers, in an effort to create as much space as possible between potential enemies and the "huddled defensive community" in the national heartland. In Soviet times, the desirability of a defensive glacis has hardly diminished in the thinking of the defense planners of this huge empire; the continuing importance of Eastern Europe in Soviet strategy testifies to this. Defensive imperatives, added to occasional surges of ambition, "fits of absent-mindedness," and attempts at tactical consolidation have produced over the centuries a Soviet-Russian empire of remarkable dimensions and diversity.

The catalog of Soviet-Russian conquest is impressive. Since the beginning of the last century Russian and Soviet forces have moved more or less successfully (and more or less permanently) into a variety of neighboring lands: the Caucasus in the early nineteenth century (Armenia, Azerbaijan, Daghestan, Georgia); the Black Sea area in the early nineteenth century (Bessarabia); central Asia in the middle and latter part of the century (Bukhara, Kazakhstan, Khiva, Kirghizia, Kokand, Tadzhikistan, Turkmenistan); the Far East in the mid-nineteenth century (the Chinese Amur regions and the Chinese Maritime Provinces); the Baltic region in 1939–40 (Estonia, Finland, Latvia, Lithuania, Poland—yet again); Eastern Europe at the end of the Second World War (Austria, Bulgaria, Czechoslovakia, East Germany, Hungary, Poland, Romania); the Far East and central Asia in the interwar years and the Second World War (the Kurile Islands, Manchuria, North Korea, Outer Mongolia, Tannu Tuva); Eastern Europe since the mid-1950s (Czechoslovakia, Hungary); and southwest Asia since the late 1970s (Afghanistan). Some of these actions were invasions; in other cases they were counterinvasions. Some were interventions; others were preventive actions. Occasionally Soviet forces withdrew, but often they did not. The size of the Soviet Union—the subject of such official pride—is not an accident of geography. It is the product of military power.

Historically, the major military problem for both tsarist and Soviet Russia has been protecting its exposed Western frontier against invasion by countries in Scandinavia and central Europe, which were technologically and militarily better developed. But with the rise of Japan and then China, major military problems reappeared on the eastern plains after a six-hundred-year break.

The historic Russian weaknesses have been: the lack of easily defensible frontiers in some regions and immensely long borders to defend; an inefficient administrative infrastructure; inadequate transport arrangements, made worse by the vast distances involved; lack of industrial development; technical backwardness; sluggish government; supply problems; and the low educational standards of military personnel. Set against such deficiencies, the country has had some enduring strengths: the vast space has been an asset in war as well as a disadvantage (as has "General Winter"); there have been abundant natural resources and great manpower reserves; brave and tough soldiers, generally prepared to accept harsh discipline and high casualties in defense of the homeland; some bright commanders; and a national commitment to military duty. Some of the worst deficiencies have been overcome in the Soviet period, but all the strengths remain. As in other aspects of Soviet life there has been a marked continuity with the tsarist experience.[8] The geopolitical and national inheritance exercises a powerful influence.

The year 1917 was in fact less significant in a military sense than some hoped and others feared. A "revolutionary" element in military policy was prominent in organization, doctrine, and ambition during the short period of Bolshevik innocence, which lasted until about 1921, but gradually many traditional elements in military behavior reasserted themselves. While the ideological element struggled to survive, the exigencies of the moment, mainly the need to overcome the challenge of the civil war, resulted in increasing weight being given to traditional professional military practices in the modernizing Red Army. The Bolshevik leaders prudently realized that they needed experts rather than Reds to run their forces if their revolution was to survive. Consequently it was not long before former Imperial officers began to reappear in uniform. It was an uneasy relationship between the Bolsheviks and the former Imperial officers, but out of the intermittent struggles between the "professionals" and the "party doctrinaires" there eventually emerged a fusion of the two heritages.[9] This was marked in 1946 with the adoption of the new title Soviet Army to replace the existing Workers' and Peasants' Red Army.

## *Strategic Policy under Stalin and Khrushchev*

Soviet leaders, like their tsarist predecessors, have never been relaxed about their state's security. The civil war (1917–20), the foreign interventions (1918–22), the period of "capitalist encirclement" when the USSR was internally weak, the Nazi threat, Japanese pressure in the Far East in the 1930s, the horrors of the Great Patriotic War, the U.S. atomic monopoly, the formation of NATO, West German rearmament, growing U.S. superiority in intercontinental delivery systems until the mid-1960s, fear of surprise attack, the dangers of polycentrism in Eastern Europe, concern about nuclear proliferation (especially by West Germany), the growing danger from China, continuing weapons modernization by NATO, the "Reagan resurgence" in U.S. military policy—all these and more, in less than seventy years, have confirmed deep national and ideological predispositions toward insecurity. The task of a Soviet military planner, who as part of his responsibility has to assume at least "fairly bad" cases when speculating about future military contingencies, has never been an easy one. Worst cases have been nightmares.

Since 1945 Soviet strategic policy has been dominated by the need to meet the requirements of deterrence and defense against the dominant U.S. enemy, although there have been some other important threats to Soviet security. In addition to reacting to these strictly "defensive" problems Soviet military strength has also been used in a variety of ways to support the advancement of foreign policy. The framework of Soviet military policy has been primarily established by defensive imperatives—rational responses to adversary challenges and technological developments—but the muscle that has been created has been used prudently to advance Soviet interests in different directions.[10]

The great victories of the Red Army in 1945 promised to solve the old problem of having to deal with incursions from the West. Not only had Germany been crushed, but the Soviet Union had been able to occupy a belt of states in Eastern Europe that would serve as a forward defense zone. Unfortunately for the Soviet leaders, the full promise of this achievement was short-lived, for their new defensive position was soon outflanked and was threatened by a greater challenge than ever before, that of the United States, with its massive air and naval power, including its monopoly of atomic weapons. However mighty the Soviet Army was, and however impenetrable its forward perimeter might become, the Soviet heartland would henceforth be exposed to the possibility of cataclysmic destruction from the air. As the cold war intensified, Soviet problems grew: the formation of NATO and the rearmament of West Germany represented a fearsome consolidation of adversary military power. As on other occasions, it was the Soviet Union's own aggressive behavior that had played a large part in provoking this increased threat to themselves. To an even greater extent than some of the leaders of their major adversary, the United States, Soviet policymakers have proved incapable of accurately understanding how they are perceived through the eyes of others.

Although Stalin still faced major strategic problems in the early 1950s, from his own paranoid perspective he must have thought that he had done a satisfactory job. He had deterred U.S. atomic-backed power and had strengthened the general defensive position of the USSR. In 1949 the Soviet Union had exploded its own atomic bomb, and the creation of pro-Soviet regimes in Eastern Europe had consolidated the USSR's forward defense zone. Soviet forces were deployed in Eastern Europe to control internal developments in the region and to extend Soviet deterrent and defensive capabilities. In particular, Soviet forces were deployed with such offensive potential that Western Europe became an explicit hostage to the USSR in the event of war. A situation of deterrence was thereby achieved by an asymmetrical balance between the U.S. lead in atomic strike power and Soviet conventional strength on the ground. But the latter was not as strong as it seemed, particularly in the early cold war period. Soviet planners themselves must have been well aware of their country's vulnerabilities. As a result, despite the devastation suffered by the Soviet Union during the war and the desperate need for reconstruction in all areas of life, Stalin nevertheless decided that the accumulation of strategic power must be a national priority. Security, as ever, was the dominant Soviet preoccupation.

Stalin demanded heavy sacrifices in order to release the resources necessary to defend the homeland against attack from land, sea, or air. In the background Soviet research and development in modern weapons was pushed along at a pace that surprised many in the West, and it soon produced an A-bomb, long-range aircraft, and different classes of tactical and strategic missiles. By the mid-1950s the U.S. homeland was, for the first time, vulnerable to massive destruction. While Soviet military doctrine appeared to be frozen by Stalinist orthodoxy ("the permanently operating factors"), in practice the assimilation of the military-technical revolution was proceeding rapidly.

Soviet strategic planning in the immediate postwar years was dominated by the possibility of war with the United States. At the same time Soviet foreign policy exhibited such signs of bellicosity that it gave the weakened and war-weary countries of Western Europe plenty of reason to fear a "march to the Channel." This fear, which stimulated the creation of NATO and greater Western military efforts, is now generally believed to have been much exaggerated. Whatever Stalin's dearest ambitions about extending the camp of socialism, his cold war bellicosity was probably a manifestation of weakness rather than a portent of aggression. The Soviet Union at this time was something of a sheep in wolf's clothing. Nevertheless, Stalin's crude addition to power and its accumulation provoked a rational suspicion in the West; there was an understandable fear of his motives and aggressiveness. When Stalin did use the military instrument to advance his foreign policy, however, he did so in a cautious manner. The consolidation of the Soviet bloc in Eastern Europe involved few military risks. Elsewhere there were a few fluid situations, and Stalin was tempted to probe to see where the boundaries of the postwar world might settle. With varying degrees of involvement, the military instrument was used to support policy toward

Iran, Greece, Turkey, Sinkiang, Berlin, and (by proxy) Korea. Soviet policy in each case was careful; and in each case Stalin was ready to accept failure.[11]

Stalin's death in 1953 had many repercussions throughout the Communist world, and not the least in Soviet military policy. A defense debate ensued regarding the character of modern war and the appropriate strategic policy for the Soviet Union.[12] The eventual outcome was the result of the interplay between factional disputes, perceptions of U.S. policy, technological dynamism, developing doctrine, and interservice politics.

Although the U.S. atomic monopoly had been broken, Soviet military planners in the middle and late 1950s still faced the problem of a possible preventive or preemptive attack. At this time the United States had great superiority in airpower, while its own homeland was still largely invulnerable to Soviet attack. Consequently, great efforts were made to improve Soviet air defense and defense against a strike from the sea. Long-range bombers and ICBMs were developed to deter attack and to retaliate if war broke out. Meanwhile Soviet ground forces in Eastern Europe and Western Russia remained deployed to keep Europe a hostage. Their training and equipment was improved as they prepared to fight a nuclear war if necessary. In 1955 the Warsaw Pact was created for a mixture of political and military reasons, including the need to rationalize military preparations for coalition warfare in modern conditions. Extrapolating into the nuclear age the old military maxim that attack is the best form of defense, Soviet forces were positioned to take the initiative at the outbreak of war and to conduct a mobile and hard-hitting campaign aimed at occupying Western Europe.

In 1960, partly because of economic strains and partly because of his growing confidence* in the stability of the balance of terror,[13] Khrushchev amended the traditional emphasis on the ground forces and instead moved toward a more ostensibly pronuclear defense posture. This has been characterized as a strategy of "minimum deterrence," but Khrushchev's domestic critics later called it "nuclearmania." Khrushchev himself had argued that if nuclear war broke out it would be settled within a few days. Priority resources should therefore be diverted to the forces that would play a dominating role in that exchange (that is, strategic offensive forces and air defense). Other forces could be reduced, because of their secondary military role and because manpower and other shortages in the economy placed a premium on cutting labor-intensive services and "metal-eaters." If the reforms had been

*This had already been manifest in 1956 when Khrushchev revised Lenin's doctrine about the inevitability of war between the two systems. Khrushchev declared that "war is not fatalistically inevitable. Today there are mighty social and political forces possessing formidable means to prevent the imperialists from unleashing war."

implemented, parts of the armed forces would have faced a hefty one-third cut. Not surprisingly, Khrushchev's axe was very unpopular among the military establishment and its supporters. Financial cuts, large-scale dismissals ("the cold purge"), and changed mission structures threatened important vested interests within the military profession, while the "one variant" doctrinal emphasis was considered to involve a dangerous inflexibility that was alien to Soviet military ways of thinking. But more pressing than these internal considerations in the minds of the critics—and eventually of Khrushchev himself—was the resurgent defense policy of the incoming Kennedy administration in the United States, which threatened to build up a dangerous lead in U.S. nuclear power. The arguments of Khrushchev's critics were greatly strengthened by Kennedy's policies and rhetoric; within a short time Khrushchev had to reverse his attempt to move the Soviet Union away from its traditionalist attitudes toward a more Western conception of mutual deterrence.[14]

The military establishment had other reasons for dissatisfaction with Khrushchev's strategic policy. In addition to what they regarded as his unsatisfactory doctrinal shift and its accompanying cuts, critics were particularly anxious about his foreign policy. His detractors were later free to describe it with some accuracy as "adventurist." Khrushchev's impulsive character, his susceptibility to the temptation of quick-fix solutions, and his ambition resulted in his using Soviet military power—the image of it, if not the reality—in a highly prominent fashion to support his foreign policy aims. The alarms of the "missile gap" period were the outcome. Between 1957 and 1962 Khrushchev practiced what has been variously called "missile diplomacy," "psycho-strategic warfare," or, less grandly, "rocket-rattling."[15] Over crises in the Middle East, the Chinese offshore islands, Berlin, and Cuba, Khrushchev engaged in threatening behavior. Western resolve held in the face of his brinkmanship.

Despite the advantages the Soviet Union seemed to possess at the time, Khrushchev failed to make his country the "wave of the future" that it threatened to be. The problem for the Soviet military establishment was that Khrushchev was carrying out his crisis-prone foreign policy without full control of events (he could not be sure of U.S. reactions) and without the actual military capability his declaratory policy suggested he had. The military's sense of vulnerability was increased by the continuing strategic efforts of the United States under Kennedy. The Cuban missile crisis of October 1962 was the denouement of Khrushchev's activist strategic policy. It ended his harebrained schemes, it exposed the strategic inferiority of the Soviet Union, and it underlined the strength of U.S. conventional military power.

Khrushchev worked hard to relax international tension after the Cuban missile crisis. Arms control played a part in this, as was manifest in the Partial Test Ban Treaty of 1963. In most respects up to this point (including his ostensible commitment to national liberation struggles) Khrushchev's strategy had provoked the United States to take actions he would not have wished. In particular, the huge buildup of strategic offensive forces by the United States in the early 1960s meant that the Soviet Union, which had taken the lead in strategic missiles in 1957, had itself to undertake a major effort throughout the rest of the 1960s in order to regain some sort of recognized parity. That this was soon achieved was largely the result of the halt in the numerical production of ICBMs by the United States at the end of the decade.

As a result of its various disagreements with and worries about Khrushchev, the Soviet military establishment was not sorry to see him ousted in October 1964. Nevertheless, in some respects Khrushchev's contribution to strategic policy did have merit from the Soviet military viewpoint. He had used military aid to support foreign policy in a generally effective way. Since the first days of the Revolution Soviet leaders had offered military assistance to support the enemies of its enemies. Khrushchev had continued this, and as a result military aid had played a large part in the ending of Stalinist isolation and in the consequent expansion of Soviet involvement in the international politics of the Third World. Soviet military as well as foreign policy took on a more global character under Khrushchev. In addition to military aid, he also oversaw important developments in military procurement. However, because of the long lead times of weapons, these effects did not become visible until the middle and late 1960s. New missile programs were begun that aimed to try at least to match U.S. strategic offensive power, and considerable attention was given to air defense, including the challenging idea of ballistic missile defense. The success of the post-Khrushchev leadership in improving the strategic balance with the United States owed more to decisions made by Khrushchev (and President Johnson) than the new Soviet leaders would have been willing to admit.

## *The Brezhnev Years*

The dominating feature of the defense policy of the Brezhnev and Kosygin regime in the second half of the 1960s was the progress made toward more or less parity with the United States in strategic offensive power. Between the early 1960s and the yearly 1970s, the Soviet Union caught up with the United States in numbers of land-bassed ICMS, from a one-to-three inferiority to a slight lead. The Soviet lead was even greater in deliverable megatonnage and missile "throw-weight," though the United States retained an impressive lead in warheads and in technological expertise. The U.S. lead also included ABM technology, which had been originally pioneered by the Soviet Union. Given its inferior industrial base, the improvement in the Soviet position was notable, even though its achievement did depend on the U.S. decision mentioned earlier to replace the doctrine of nuclear superiority with that of sufficiency and parity.

The first SALT agreements of May 1972 crowned the Soviet strategic efforts of the 1960s. The Russians had moved from the position of inferiority exposed at the time of the Cuban missile crisis to one of real strength, new confidence, and recognized formal parity in offensive and defensive systems.

Alongside the growth of strategic offensive power, the second half of the 1960s also saw the accumulation of Soviet conventional military strength. Prominent in this respect was the development of greater airlift and naval capabilities. Continuous attention to Soviet air power in the 1960s resulted in a much improved capability. The growth of airlift potential was substantial and was demonstrated operationally in the reinforcement lifts to the Middle East after the 1967 and 1973 wars, and the combat landing in Czechoslovakia in 1968; in addition, airlift showed its potential in a series of large-scale exercises.[16] However, it was the developing strength of the Soviet navy that became the focus of particular Western interest at the turn of the 1960s and 1970s.[17]

The improved quality of the forward deployment of Soviet navy units was primarily motivated by the general war requirements of strategic deterrence and defense, and in particular by the need to be in a position to counter the nuclear threat represented by strike carriers and Polaris and Poseidon submarines. The story of this effort through the 1960s and 1970s was one of Soviet progress being outflanked by the improving technology of the U.S. Navy. As far as ballistic missile-firing submarines (SSBNs) are concerned, this is still the case, with Trident adding complexity to an already immense problem. But it has not been as easy for the strike carriers of the U.S. Navy to outflank their Soviet shadowers, and their vulnerability has increased. In the late 1960s the strategic strike mission became more important for the Soviet navy itself, and with the deployment of its Delta-class SSBN after 1973, it became possible for the USSR to threaten to strike the U.S. homeland from waters contiguous to the Soviet Union. As a result of this, and of Soviet worries about Western capabilities in antisubmarine warfare (ASW), the Soviets assigned more surface forces to the active defense of their missile-firing submarines (a mission that has been dubbed "pro-SSBN"). From "protected ocean bastions" Soviet submarines can strike the United States or Western Europe or withhold strike for the purpose of intrawar deterrence and bargaining in the event of a prolonged conflict. The pro-SSBN mission demands a heavy investment of forces.[18]

In addition to their general war tasks, the Soviet naval units that were drawn forward have been used on a number of occasions, and in a variety of ways, to support Soviet foreign policy in low-level, nonrisk activities off the coasts of West and East Africa, in the Persian Gulf, and during crises in the Middle East and the Indian subcontinent.[19] The new strength and activities of the Soviet navy markedly complicated Western naval planning, but serious weaknesses remained in the Soviet navy's capability throughout the 1970s. In particular, it could not launch and sustain a large intervention. Related to this were its shortcomings in organic air power.

A major effort to remedy some of these weaknesses began in the late 1970s, when a marked increase took place in the resources allocated to warship building.[20] New classes of more capable ships were planned, and follow-on classes were to be significantly bigger or better, with more endurance and larger weapon loads. The outcome of this increased naval capability will, in due course, be a significant improvement in Soviet naval airpower (land-based in the form of the Backfire bomber designated for the navy, and sea-based as a result of the embryonic carrier development[21]), better underway replenishment, more effective measures to counter Western submarine strike forces, and additional general-purpose forces. Of these developments the most eye-catching was the decision to build a number of large air-superiority carriers. Behind all this new hardware lay the increased authority of Admiral Gorshkov, the commander in chief of the Soviet Navy since 1956. In a series of famous articles in the early 1970s, later developed into a book, Gorshkov argued successfully against the army-dominated system in favor of the importance of balanced seapower in peace as well as war.[22]

By the late 1980s the Soviet navy will have an even more capable force than it had in the 1970s. But it remains to be seen whether this capability will translate into real power—the ability to encourage others to behave in a desired fashion. In part this will depend upon the Western response. For the moment, as Michael MccGwire has put it, "The present naval posture is one of being on guard against the possibility of war, rather than preparing for its inevitability; of being prepositioned rather than poised." Whatever the Soviet motives, the capability is being put together for a new instrument of overseas policy.[23] Its uses and utility remain to be tested.

While the Soviet navy achieved a new significance after the mid-1960s, the ground forces reestablished some of the traditional importance they had lost under Khrushchev. There was a pronounced increase in their capabilities in both Eastern Europe and the Soviet Far East, where the confrontation with China became increasingly serious at the turn of the 1960s and 1970s. The Soviet forces in both areas developed more striking power and a greater flexibility in doctrine.

Under Brezhnev Soviet forces in Eastern Europe continued to provide a range of defensive and offensive capabilities in the anti-NATO context, and also contributed through their looming military might to the support of Soviet diplomatic positions. Soviet military doctrine adjusted to the possibility of prolonged conventional operations at the outbreak of a European war, partly in response to NATO's shift to so-called flexible response in 1967, and partly because of the Soviet Union's own politico-military requirements. In particular, Soviet forces in Eastern Europe remained a major factor in the maintenance of Soviet control over the region, by latent or actual threat, including the threat of "mobilization by manoeuvre."[24] Soviet military might, and the demonstrated will to use it, has established limits to the independent aspirations of Eastern European countries. This was underlined by the invasion of Czechoslovakia in 1968, and yet again during the simmering Polish crisis of the late 1970s and early 1980s. Soviet military force was not only the backdrop and ultimate instrument of policy in Eastern Europe, it was also both backdrop and increasingly useful tool for advancing a range of Soviet interests elsewhere.[25]

In general, Soviet behavior has been cautious when projecting military power outside its borders. Ever since the Bolsheviks' failure in Poland in 1920–21, up until the intervention in Afghanistan in 1979, Soviet troops have not normally been used directly to try to spread the area under Moscow's control.[26] There have been two exceptions to this operational principle first established by Lenin, and both belong to very special circumstances. First, there was the prelude to the Second World War, when the Soviet Union embarked upon preventive expansion into Poland, the Baltic Republics, and Finland; and second, at the end of the war there was the defensive-ambitious policy of staying where the counteroffensive against Germany had taken Soviet forces. The Leninist principle (nothing if not pragmatic during the years of Soviet military vulnerability) was questioned in an internal debate in the 1969–71 period, when an overseas role was contemplated for Soviet forces.[27] In the end what emerged was an agreement to further the Soviet Union's "internationalist mission" (supporting "national liberation movements," countering "imperialism," and so on) but without using Soviet troops directly. The Soviet Union would provide advisers, military aid, and logistical support, while the fighting would be done by others. In the event the Cubans bore the brunt. As a result—notably in Angola and Ethiopia—the Soviet Union got the best of both worlds. In one case a national liberation movement was helped to power; and in the other a Marxist regime was kept in power. In both cases these successes were achieved at acceptable costs.

The culmination of the growing Soviet military capability and their apparent confidence in using it occurred at the end of 1979, with the intervention in Afghanistan. For the first time since before the war against Germany, the Soviet Union directly injected large numbers of its troops in a country outside the Soviet bloc. While the precise motives for the intervention in Afghanistan remain the subject of much discussion,[28] the circumstances were clear. In the background the Soviet Union was getting less and less out of what was rather erroneously still labeled its détente relationship with the United States, while in the foreground 1979 saw a sharp deterioration in the position of the pro-Soviet Marxist-Leninist government that had seized power in Kabul the year before. For a time the Soviet Union tried to deal with the problem by political means (including intrigue) and military aid, but in the end it decided to send in its own forces rather than suffer the loss of a pro-Soviet government on its southern borders. Such a loss might have led to prolonged instability in the area and possibly the creation of an opportunity for increased meddling on the part of a more activist and interventionist United States.[29] So far, Soviet forces have failed to impose their authority over the fiercely independent Afghans, and the intervention has led to a deterioration in Soviet relations with the West, China, and many Third World countries. The one military adventure beyond the Soviet bloc is proving to be a heavy burden in both military and political terms.

In trying to explain the growth of Soviet military power through the 1960s and 1970s, which gave its possessors many more options and a degree more confidence in employment, it is tempting to impose a coherence on the process that was probably not there in practice.[30] Rather than the buildup representing the working out of a blueprint, the product of a single decision, it is more likely that it was incremental and evolutionary, the outcome of a series of specific responses to developments elsewhere, especially in the United States and China. The rapid expansion of the Strategic Rocket Forces after 1964, for example, was the result of decisions taken in 1961 in reaction to the huge increase in U.S. strategic strike promised by the Kennedy-McNamara drive for superiority.[31] The moving forward into the open oceans by the Soviet navy on a semipermanent and then permanent basis, first evident in the mid-1960s, was the result of decisions taken in the late 1950s and early 1960s (that is, predating the Cuban missile crisis), in order to meet the U.S. strategic strike threat at sea, in the shape of carriers and then Polaris.[32] The revival of the ground forces under Brezhnev cannot be explained as neatly in terms of a specific response to a specific problem. Instead it sprang from a variety of internal and external developments, notably the decision to reverse Khrushchev's 1960 cuts, the willingness of the Brezhnev-Kosygin leadership to defer to army advice and demands, and NATO's shift to "flexible response," which in turn led to new Soviet interest in the prospect of prolonged conventional operations at the start of any war. This revival of the ground forces, which was accompanied by the development of new weapons, a steady increase in conventional armaments stocks, and the deployment of new theater nuclear weapons, all became more urgent in the wake of the serious deterioration of Soviet relations with China following their limited border hostilities in 1969. As will be discussed later, by the second half of the 1960s, war with China was the most likely if not the ultimate threat for Soviet military planners.

It would be foolish, of course, to try to explain anything in strategic life with reference to a single simple cause. The Soviet air force, for example, has been one Soviet service whose growth cannot easily be explained in terms of a response to turning points in Western policy. Rather, Soviet air defense, lift, reconnaissance, and strike all received continuous attention through the 1960s and reflected a consistent recognition by the Soviet command of the potentialities of air power.[33] The Soviet air forces apart, the reactive and incrementalist interpretation of Soviet behavior explains the general thrust of the buildup, while the details were the result of technological constraints, economic pressures, military traditions, interservice rivalry, institutional traditions—and all the other factors to which the development of armed forces are prey. Whatever the particular combination of motives that led to the Soviet military build-up, the fact is that the Soviet leaders have possessed since the late 1960s a modern and impressive military machine; and this has provided them with options they did not formerly have.

The idea that the Soviet military buildup of the 1960s and 1970s was largely a reaction to developments elsewhere is not readily accepted by many Western observers. The majority of the latter see the strategic confrontation entirely in terms of Western responses to Soviet challenges. But the contrary (Soviet) viewpoint is legitimate, and if one fails to give it due recognition, then an important element in our strategic predicament will have been missed. This point is well illustrated by the clash of interpretations over the deterioration of détente in the 1970s.

It is almost an article of faith with Western observers, especially in the United States, that détente collapsed as a result of Soviet behavior. They point to the inexorable Soviet arms buildup, Soviet adventurism in the Third World, and Soviet obstructionism on the subject of arms control. The view from the Kremlin, however, even when ritual demonology and instinctive propaganda are taken into account, is justifiably different. In the first place, the Russians believe that their own arms buildup was not inconsistent with détente, but was instead a necessary step to create the military stability that was the foundation for a relaxation of tensions between the superpowers. In addition, they have argued that their support of national liberation movements was both a duty and a legitimate act of assistance. Secondly, the Kremlin interpretation of events suggests that it was the U.S. arms buildup through the 1970s, uncooperative U.S. behavior under the label détente, and the lack of seriousness on the part of U.S. administrations regarding arms control that ultimately brought superpower relations to their worst state since the Cuban missile crisis. From the Soviet viewpoint, the 1970s was characterized by a long series of hostile U.S. actions. There was the development and deployment of major new weapons systems (Trident, MX,

Pershing II, the B-1 bomber, long-range cruise missiles). There were changes in U.S. strategic doctrine that worried the Soviets (e.g., talk of limited nuclear options and the shift to war fighting). There was unfriendly behavior in relation to bilateral agreements—the Americans failed to follow through on economic deals and squeezed the Soviet Union out of Middle Eastern affairs. And there were hostile diplomatic positions—the United States pressed human rights issues, engaged in tough talk and actions over Afghanistan and the Polish crisis, normalized relations with China, and encouraged Japan to take on a bigger military role. Finally, the Soviets criticized U.S. arms control positions (failing to ratify SALT II, the uncompromising U.S. attitudes on a range of other talks from naval restraint in the Indian Ocean to banning antisatellite weapons). With this image of American behavior during détente, the Russians felt that they were justified in responding toughly, cautiously, and suspiciously themselves. "Needless to say," as Marshall Shulman has written, "the Soviet leadership perceived these [American] actions not as responses to Soviet behavior but as indications of a militant American intent."[34]

As a result of its extensive military efforts over two decades, the Soviet Union—long accustomed to be number two in the arms race[35]—caught up with the United States under Brezhnev's leadership. By the early 1980s, the two powers reached what Zbigniew Brzezinski has felicitously called an "ambiguous equivalence." It is possible to argue the numbers game regarding the military balance until one is blue—or red—in the face. Although the numerical strengths and weaknesses of the superpowers can be manipulated in a variety of ways, uncommitted observers are confident that neither has any meaningful "superiority," either at the nuclear level or lower down. Even with conventional forces it has become apparent over the years that Soviet strength has been consistently exaggerated, and that the gap between NATO and the Warsaw Pact is not as enormous as has sometimes been suggested.[36]

Despite the strains on Soviet defense policy in Brezhnev's later years as a result of the resurgence of American defense efforts under President Reagan, the position of Soviet forces at the time of Brezhnev's death in 1982 was generally satisfactory to his military establishment. At the level of general-purpose forces Soviet capabilities and doctrine had become more well rounded. At the level of theater nuclear forces in Europe and the Far East older inadequate systems had begun to be replaced by much more effective modern ones. At the strategic level the inferiority of the Khrushchev period had been overcome and formal parity had been achieved. In addition, this military buildup had been allied to a policy of peaceful coexistence that had been much less risk-prone than under Khrushchev, but that had managed on several occasions to advance Soviet interests in parts of the Third World. One consequence was that relations between the party and the military establishment improved markedly during the Brezhnev years when compared with the final stages of the Khrushchev period.[37]

## *Defense Decision-Making*

As was apparent during the Khrushchev years, relations between the top military and political leaders are at the nub of the secretive process by which defense decisions are made in the Soviet Union. Although detailed knowledge about this process is lacking, three general characteristics seem evident. First, the process since the mid-1950s has become more bureaucratic than in Stalin's time; second, the military role in policy-making has in some cases increased; and third, there is not an inevitable tension between the party and military establishments, even on resource allocation questions.

The Soviet tradition in defense policy-making has been one of firm civilian (party) control; indeed, there has been a recurring fear of undue military influence, or "Bonapartism"—the fear of a charismatic military leader who will pose a challenge to the established order. As a mark of this the military establishment has normally been formally excluded from decision-making power, though certainly not close consultation. At the pinnacle of the Soviet decision-making hierarchy is the Politburo, which, in its long history, has not usually included the minister of defense; however, this has generally changed since 1973, when Marshal Grechko was appointed to that body, as, subsequently, was his successor, Marshal Ustinov. But when Ustinov died in 1984 the Politburo reverted to its former practice. The new defense minister, Marshal Sokolov, was not one of those promoted to the Politburo following Mikhail Gorbachev's accession to power in March 1985; he remained a candidate or nonvoting member. This was but one of several signs that suggested that the new general secretary may have wanted to play down the prominence of the military. Nobody could say for sure whether this meant that Gorbachev was less dependent on the backing of the military establishment than his predecessors as general secretary.

Theory and practice suggest that the party leaders in the Politburo have reserved for themselves the right to make decisions on the employment of forces and on questions of peace and war.[38] Other military matters are formally passed by the political leadership to the Ministry of Defense, where they are then dealt with by the appropriate body, be it the General Staff, the Military Academies, or the Chief Political Directorate of the Armed Forces. Military doctrine (such questions as future types of war or weapon employment) is evolved in these bodies.

Linking the political and military hierarchies are several joint committees, in which details of policy are probably worked out. By far the most important of these bodies is the Defense Council, whose existence was only revealed in public in 1976, though its importance had been suspected for some years by Western analysts. Great uncertainty surrounds the activities of the Defense Council, but what is known points to the fact that it is the foremost Soviet decision-making body for national security affairs, whether or not the country is at war.[39] It is made up of a small group within the Politburo and includes the minister of defense and other military and political leaders as required; it is chaired by the general secretary of the party.

In general, political-military consultation at the highest levels in the Soviet Union is thought to be highly regular, with business being cyclical and mainly concerned with incremental decisions and with an agenda that is probably strictly confined to the military field. In the Defense Council, as in other bodies, the party leadership makes clear its authority in defense matters, but this does not rule out the possibility that the military establishment will sometimes exert itself—acceptably or unacceptably—to influence policy. Many observers speculated that clashes over military policy led in September 1984 to the apparent dismissal of Marshal Nikolai Ogarkov from the post of chief of the Soviet General Staff. However, Ogarkov was not completely stripped of power; he was instead moved to the headquarters of the western military district at Minsk, and his name continued to appear periodically on important occasions. The curious episode took a further twist in mid-1985 when the Ministry of Defense published a short book by Ogarkov, *History Teaches Vigilance*, in which he stressed the dangers of ignoring new weapons technology and criticized the pointless accumulation of strategic nuclear weapons. Indeed, hindsight now suggests that his dismissal was, in fact, a sideways promotion.

The formal position, therefore, is one of firm civilian control. However, this should not be taken to mean that the military establishment is not able to affect the thinking of the party leadership. Indeed, in some respects its role has grown since the late 1960s. This latter trend is not the result of a drop in traditional party vigilance or of irresistible military pressure, although it may have been encouraged by an aging and more collective party leadership and a military establishment with some self-confident characters. The main factor behind the growth in importance of the military establishment in Soviet policy-making has been the thrust of events; over the last two decades the character of certain issues has compelled the Soviet leaders to seek the advice of their military specialists. The invasion of Czechoslovakia, the confrontation with China, SALT, the continuing Middle East crisis, military aid diplomacy, forward naval deployment, the mutual balanced force reduction talks (MBFR), the Angolan crisis, the trouble in the Horn of Africa, the intervention in Afghanistan, NATO's decision to deploy cruise missiles and Pershing, the intermediate nuclear force (INF) talks, the expanded U.S. defense budget—all these issues (and more) have required specialist military advice and have involved party-military consultation at the highest level.

Events have therefore sucked a less than bashful military establishment into the decision-making process, and this has been encouraged by leaders with the bureaucrat's tendency to seek advice and spread responsibility. The latter was a notable feature of the Brezhnev leadership in comparison with those of his predecessors; and the military establishment, with its memory of the problems of the Khrushchev years, did not hesitate to accept the increased advisory role. In some circumstances the military contribution may go beyond advice. When there is a relatively weak Soviet leader—as seemed to be the case under Konstantin Chernenko—the opportunity was given for strong Politburo members like Andrei Gromyko and Dimitri Ustinov to gain more authority in their own professional spheres; that is, foreign and defense policy respectively. In Ustinov's case it might be assumed that the more autonomy and authority he gained the greater the opportunity for the military establishment to increase its influence. This does not necessarily follow. In practice it is always difficult to guess the extent to which a Soviet minister of defense will act as a party man or as a spokesman for the professional military viewpoint. Presumably Grechko was more a spokesman of the latter than Ustinov, since Ustinov's earlier career involved running the defense industries rather than being a professional soldier.

The informal aspects of the party-military relationship tend to become important when the military establishment fails to achieve its objectives through the formal channels, or when it wishes to protest against decisions it regards as threatening to its interests.[40] Various opportunities can then be exploited. For example, the military may cultivate personal relations with individual political leaders; the latter, for obvious reasons of their own, are usually interested in developing harmonious relations with military leaders. Alternatively, the military establishment might take its case to wider party and government opinion, in the hope of affecting decisions from below. The most notable examples of such behavior include the military establishment's opposition to Khrushchev's proposed force cuts in the early 1960s, its attempts to influence budgetary decisions in 1966 and 1967, its skepticism about arms control (including initial opposition to SALT), and its unease about détente in the 1970s. In 1984, as was suggested earlier, clashes of opinion with the Kremlin over various aspects of military policy may have been the reason for the dismissal from his post of Chief of the General Staff of Marshal Ogarkov, reputedly the cleverest soldier of his generation.

Although there have been important differences of opinion from time to time between the Soviet political and military hierarchies—indeed there is always likely to be a degree of tension in the relationship between a professional officer corps and a totalitarian party[41]—there remains a vital area of interdependence. Because of the party's traditional preoccupation with security, the need to have military backing for policies, and the increased military content of important areas of foreign policy, the party is both reliant on and susceptible to military advice. As a result, it is not surprising that the Soviet military establishment, in the view of one close observer, "is consulted at an earlier stage and in greater depth on foreign and defence policy than ever before."[42] Depite this, however, there seems little doubt that on all major matters of military policy the party has, and intends to keep, the final word. Overambitious, interfering, or exceptionally clever military commanders—or any combination of these qualities—can expect trouble. The careers of Tukachevski in the 1930s, Zhukov in the 1950s, and Ogarkov in the 1980s testify to this.

While occasional periods of tension can be expected between the party and military hierarchies, there is a basic interdependence between them. The military looks to the Politburo for backing for its preferred programs and the expenditure to support them, while the party leaders look to the military for support in struggles for succession and in subsequent efforts to maintain authority. The military establishment belongs to that trinity of organizations, together with the party apparatus and the KGB, that every new Soviet leader must attempt to satisfy. Meeting the interests of the military is always likely to be particularly important for those Soviet leaders whose own rise to supreme power has in part depended upon the support of the military establishment; this was the case with Leonid Brezhnev and Yuri Andropov.

In general, if not in particular, satisfying military needs will not be something security-conscious Soviet leaders will find uncongenial. One does not become general secretary of the Communist Party of the Soviet Union by having dovish sentiments or by being seen to be squeamish when it comes to making hard choices about military matters. There is therefore an important overlapping of outlooks as well as interests between the two groups. The Soviet "civilian" leadership could well be the most security-conscious and military-minded of any in the developed world; it is therefore the one most likely to give a sympathetic ear to the military case. As William Odom has put it, in Moscow "martial preferences are [not] limited to marshals."[43] This sharing of interests, attitudes, and personnel therefore does not make it profitable to dwell too long on the issue of the military influence on policy-making in the Soviet Union. As it is sometimes put: "There is not a military-industrial complex in the Soviet Union; the whole country is a military-industrial complex."

While Soviet leaders since Stalin have become more willing to listen to the professionals in the making of military policy, the range of advice on which they can call is markedly narrower than that of Western governments. In the Soviet Union there simply does not exist the public debating of defense issues that characterizes the Western world in the press, radio and TV, in parliament, or in specialist journals. The Soviet Union lacks the phenomenon of independent academic strategists in universities, or civilian specialists in research institutes, or unofficial newspaper defense correspondents. Information about defense is guarded carefully by the military.* But this does not mean that the phenomenon of defense debates is entirely absent; as earlier discussion has indicated, there have been both differences of opinion about defense and outlets for its expression. In fact, in the postwar years there have been several periods of vigorous discussion about military matters, notably over resource allocation questions, the role of nuclear weapons, the future of naval strategy, and the overseas role of Soviet forces. As the debating occurs in a distinctive political and cultural milieu, it is hardly surprising that the activity is different from the Washington model. The debates are more cautious, they are technically rather than politically inclined, and outsiders have no role. Military affairs are dealt with by a smaller, more inbred, and more professionally involved group than in the West.

While ultimate authority on defense matters resides in the Politburo, the formulation of military policy in the Soviet Union is always likely to be much more complex than administrative flowcharts or simple decision-making models might suggest. The policy process, always complex enough, has become much more so than in Stalin's time when the system was dominated by one man, and when clearer military imperatives resulted in requirements-led innovation. Since the mid-1950s, and especially since Khrushchev's fall, various aspects of bureaucratic politics have come to the fore. There has been more formalism, greater bureaucratic activism, increased professional expertise, newer techniques in decision making, and a diffusion of responsibility on nonvital matters. Furthermore, interest-group activity has been greater than was conceivable in Stalin's time. In weapons procurement the relevant ministries, scientific research institutes, and design bureaus bring influence to bear, and there is no lack of interservice rivalry and competition within the traditionally army-dominated system. "Weapons dynamics" and economic factors also have their pull on the procurement process.[44] The increasing complexity of the factors involved adds considerably to the problems of policy-making in this difficult area.

Looking to the future, it might be expected that the increased bureaucratization of the military decision-making process will make policy

*The standard illustration of this is drawn from the SALT I negotiations, at which one-third of the Soviet delegates were military men, including the second-ranked member of the negotiating team, Colonel-General Ogarkov. At one point Ogarkov criticized U.S. negotiators for giving the civilian members of his delegation information about the Soviet armed forces that he felt Soviet civilians should not possess. John Newhouse, *The Story of SALT* (New York: Holt, Rinehart and Winston, 1973), 56.

more expert, more incremental, and less susceptible to one man's proclivity to harebrained schemes. At the same time it could mean that mistakes will be carried through with ever more efficiency.

## *Military Policy in the Domestic Environment*

One area in which the party and the military establishment have clashed in the past, and might again in the future, is in the economic dimension of military policy. This clash is familiar to all political systems. In the Soviet case care must be taken against overdrawing the economic problems involved and the extent of the differences between the party and military establishments.

Calculating Soviet defense expenditure is a perplexing task, and one that adds to the difficulty of determining the burden of defense on the Soviet system.[45] Western estimates of Soviet defense spending vary, but it is usually reckoned to be about 12 to 16 percent of the GNP. In contrast, Soviet authorities claim that only 3 percent of their GNP goes to defense. Making the numbers add up and then mean something in Western terms presents defense economists with a major riddle, as a result of the different accounting systems, costing methods, and so on.

Whatever the amount of money actually spent on defense in the Soviet Union, Western observers agreed in the late 1970s and early 1980s that it represented an increasingly heavy burden on the economy, and therefore a growing political headache. It was usually assumed that defense costs were growing at a constant 4 to 5 percent a year throughout the 1970s and that they would continue to do so. This would mean that the military burden would become ever more pressing, since the economy as a whole was only growing at 2 percent. At the end of 1983, however, the CIA revised its calculations about Soviet defense spending and produced conclusions that were a surprise to most observers—so much so that they have still not been assimilated by those who always prefer to inflate the Soviet military threat.

The revised CIA estimates show that Soviet military spending was consistently overestimated through the latter part of the 1970s.[46] Despite the picture that had been widely presented, and believed, of a relentless Soviet arms buildup, what emerged from the revised estimates was somewhat different:

- The Soviet military buildup in the 1960s and 1970s actually slowed down significantly after 1976.
- Between 1977 and 1981 Soviet procurement of new weapons measured in dollar terms did not grow at all.
- The total of Soviet defense spending slowed down markedly after 1976. For the previous decade it had grown at constant rates of nearly 5

percent per annum, measured in rubles, but it declined to only 2 percent after 1976. The latter meant that the growth in defense spending was not outstripping growth in the economy as a whole, but was equivalent to it.

• The slowdown in defense cost growth rates was distributed unevenly, but between 1977 and 1981 the procurement of military hardware did not grow at all in dollars. Strategic offensive weapons—which had aroused such disquiet in the United States—were being produced and deployed at "relatively relaxed rates" after 1976.

• It was previously assumed that the defense share of the GNP increased from 12 or 13 percent in 1970–78 to 13 or 14 percent in 1979–81. The present view is that the share was 13 or 14 percent in 1970–78 and remained unchanged in 1979–81.

These figures, and the associated analyses, are very significant. They show that the Soviet military buildup was much slower in the 1970s than most people in the West believed, and still believe; the figures show that the growth of Soviet spending was less than that of the United States (whose economy is about twice the size of its superpower rival); the figures call into question the opinion that the Kremlin does not consider costs as a major factor in its decisions about the military; and they add further weight against those who argue that the Soviet Union runs virtually "two economies," a highly efficient military one and a backward civilian counterpart.

From the revised figures it would appear that any strains that have and will develop between the military and political establishments over the level of the defense effort will be more the result of political choices than economic imperatives. The Soviet defense burden is certainly considerable, but it is less than it is often thought to be in the West. Since Soviet defense spending has been matching economic performance rather than outrunning it in recent years, there seems little prospect, as the American Right had misguidedly hoped, that economic strain will lead to a significant reduction in Soviet military effort. Equally, the revised figures mean that the Russians are not likely to be driven into arms control, as some Western liberals have fondly imagined. Having said that, one should note the Soviet leadership under Gorbachev would presumably prefer to spend less on defense—if the international situation allows it—so that the country's disappointing economic prospects can be improved. During his first months in power, Gorbachev made it apparent that better economic performance was a first priority.

Despite some of the preceding comments, it nevertheless remains true that the defense burden is a major problem for the Soviet leadership, as indeed it is for the governments of all the industrialized world. In the Soviet Union—as elsewhere, but perhaps to a greater extent—defense spending has led to sacrifices and distortions in the civilian sector. In the 1970s, under Brezhnev, investment in industrial development was curtailed to help meet military demands, while inefficiency in the civilian sector was encouraged by the fact that military industry had a priority in supplies and skills. This led to the familiar opinion that the Soviet leaders virtually ran two parallel economies, the military and the civilian—the one efficient and effective, with good quality control and steady innovation, but the other inefficient and backward, characterized by poor quality and shortages.[47] However, the contrast between the "two economies" should not be pushed too far, in the light of some of the problems of innovation and reliability shown in tanks, aircraft, missiles, and other aspects of the Soviet military machine, not to mention the large contribution the defense industry makes to civilian output.[48] It is also easy to exaggerate the "guns versus butter" dilemma in the Soviet Union, although it undoubtedly exists. The higher reaches of the military establishment probably realizes—or has been made to realize—that the country does face economic difficulties and that these have defense implications. This message came through clearly during Yuri Andropov's brief period as general secretary, when he tried to tackle some of the economic problems that had accumulated during the Brezhnev years. The Soviet military establishment has probably been made to appreciate, even if not fully, that the continued growth of Soviet military power depends upon a dynamic economic system. This means that the military and party leadership have a common interest in at least some degree of economic reform. Whether or not such reform will be attempted by Gorbachev, there seems little reason to suppose that Soviet military leaders will cease to press for, and receive, most of the military equipment they consider necessary for the country's security.

Without doubt, the Soviet armed forces are a serious burden on the economy, but they also perform a number of essential services in the domestic setting. As armies in many other countries do, the Soviet armed forces have important internal roles.[49] Border patrol has been a traditional function; the orientation of large numbers of Soviet troops has always been inward, to check defection, as well as outward to prevent intrusion. In addition, the armed forces have contributed directly on a number of occasions to internal security in times of crisis, notably in the suppression of several insurrections. The armed forces also undertake a variety of socializing functions; as well as increasing the country's pool of military manpower, the institution of conscription has been as an instrument of political indoctrination, nation-building, youth control, and social conformity.

For a mixture of historical and practical reasons, the relations between the civilian and military sections of the Soviet community have continued to be reasonably close, with none of the outright antimilitary attitudes that have been displayed in various Western industrialized countries. While the military profession might not today have all the prestige and attractiveness of the past, it remains an accepted and useful part of national life. Taken as a whole, the internal functions performed by the armed forces contribute in important ways to the effectiveness of the Soviet political system. These functions might well increase in importance in the years ahead as the proportion of Great Russians making up the Soviet state continues to decline and as ideological fatigue continues to grow. One is tempted to conclude that if Soviet leaders did not feel that they need an army for external reasons, they would have to invent one for internal ones.

## *Alliances and Soviet Strategy: The Warsaw Pact*

Soviet troops have had far less operational experience since 1945 than their counterparts in the armed forces of the major Western military powers. Ironically, with the exception of the intervention in Afghanistan, what operational experience they have had has been at the expense of some of their country's ostensible allies, namely, Hungary, Czechoslovakia, and China. Furthermore, mobilizations have also been undertaken against other allies, notably Romania and Poland. In Europe and elsewhere, it seems, Socialist states attack each other far more often than they do neighboring capitalist countries. With comrades like these. . . .

Eastern Europe has always been a priority for Soviet foreign and defense planners. It will be no less so in the future than in the past, though the problems of control may well multiply.[50] One reflection of this enduring Soviet concern with Eastern Europe has been the growing importance of the Warsaw Pact.

The Warsaw Treaty of Friendship, Co-operation and Mutual Assistance was signed between the Soviet Union and seven Eastern European states in May 1955.[51] It is generally agreed that its origins were primarily political rather than military. In part it was a Soviet attempt to deal with a number of immediate problems. These included the desire to offset the entry of West Germany into the Western European Union and NATO, the wish for a propaganda counter to NATO, and the need to provide a rationale for military movements in Eastern Europe after the Austrian State Treaty (which led to the withdrawal of Soviet forces from its zone of occupation in that country). Important as these reasons were, however, the pact was primarily created as a means of rationalizing and institutionalizing Soviet political and military authority in Eastern Europe on a more acceptable basis. In addition to direct Soviet purposes, the pact was also to serve—and continues to serve—some internal and external interests of some of the socialist regimes in Eastern Europe. While the latter's contribution to the alliance is not be overlooked, the evolution of the pact must be related in the main to Soviet attempts to deal with the problems of an evolving Eastern Europe in a world of change.

For its first five years, until 1960–61, the political and military activity of the Warsaw Pact was very limited. Since that time, however, a combination of factors have worked to increase its significance for its members. The first notable development was the start of a series of major joint military exercises at the start of the 1960s. But while the military side of the pact progressed through the mid-1960s, political tensions within Eastern Europe began to trouble the Soviet Union. The swing from Soviet-defined rectitude by Albania, Romania, and then Czechoslovakia each demanded attention, and the pact was used in attempts to deal with the various problems these countries created by exclusion, pressure, threat, or directive. By 1967 the pact was playing a novel role in Soviet-Romanian relations in particular; it had become an important channel for communication and conflict resolution.[52] Unfortunately for all concerned, this embryonic development was not able to contain what the Soviet leaders and their Eastern European supporters regarded as the dangerous side of Czechoslovakian "socialism with a human face."

In late August 1968 an invasion of Czechoslovakia was carried out after the failure of multilateral diplomacy backed by military demonstration. The Warsaw Pact had been prominent in the buildup of military pressure in the summer of 1968, but command of the actual invasion was handed to General Pavlovski, the commander in chief of the Soviet ground forces. This episode appeared to confirm Malcolm Mackintosh's thesis that in its military sphere the pact is not intended to be an operational command-and-control headquarters, but rather an administrative authority designed to organize and coordinate the activities of the members. It is a war department rather than a supreme headquarters.[53]

The Warsaw Pact has grown in both military and political importance since 1968. The Soviet leaders recognized that aspects of the organization needed changing, and this was done in the so-called Budapest reforms of March 1969. The reforms involved some reorganization of the military side of the alliance, including the granting of a larger formal role to the non-Soviet members. From the outset, however, the Soviet Union has had a monopoly of the senior posts within the pact, has dominated the decision-making process, and has been preponderant in all aspects of its military activities. The Budapest reforms altered none of this; the changes were largely cosmetic—but that in itself was important to the Eastern Europeans.

In addition to the manifestations of Soviet domination just mentioned, another arrangement of note has been the presence of (Soviet) Joint High Command representatives in the capitals of the other allies. Their formal task is to help ensure that training and other norms are being met, but this arrangement also provides intelligence about the allies and a further degree of control. It is one of the several means by which the pact helps to organize the armed forces of Eastern Europe in such a way as to limit their capacity to resist Soviet military pressure, either latent or actual.[54]

One outcome of the increased attention given to the pact since 1968 has been a general improvement in all aspects of its military capability. Through the 1970s manpower levels increased (from 3.7 to 4.8 million), as did the quality and quantity of tanks, self-propelled guns, battlefield missiles (surface-to-surface and surface-to-air), tactical aviation, helicopters, and transport potential (bridging and logistics). This combat potential has been regularly and effectively displayed in a series of large-scale and well-publicized exercises. Such an improvement in combat power is consistent with a doctrine that envisages the possibility of prolonged conventional operations in the opening stages of hostilities of Europe.[55] The role of the non-Soviet allies in Soviet strategy should not be minimized. It is not sufficiently realized in the West, for example, that in the event of an unreinforced attack on NATO—a scenario envisaged by some U.S. planners since the late 1970s—over half of the in-place Warsaw Pact divisions would be non-Soviet.[56] When one considers that the bulk of these forces will be Polish or Czech, the serious military implications of the internal developments of these countries becomes readily apparent. Furthermore, the combat potential of the non-Soviet forces varies considerably, though mainly between the "northern" and "southern tier" members, rather than between the "front line" northern tier allies.* Manning levels, equipment types, training, and morale present a much more complex picture than might first appear from scanning the cold statistics of *The Military Balance.* Indeed, after closely examining the different histories and characteristics of the various national armies making up the pact, one should grow accustomed to thinking twice before speaking about the forces of the "northern tier," let alone the Warsaw Pact.[57]

A crucial question facing Soviet planners concerns the reliability of its allies in the event of war. Planners can never give certain answers to such questions, but the most often heard guess in the West states that morale problems in the non-Soviet allies would grow the longer any war continued. But even in a short war the Soviet High Command would worry greatly about the fact that their major reinforcement routes have to pass through Poland. For one reason or another, if ever the Soviet command did decide to attack westward, every consideration would favor a speedy campaign.

Alongside the "perfecting" of the military organization of the pact through the 1970s, there was an upgrading of its political role. Especially in the immediate aftermath of the invasion of Czechoslovakia, this up-

---

*The northern tier includes Czechoslovakia, East Germany, and Poland. The southern tier allies are Bulgaria, Hungary, and Romania.

grading involved stressing the Socialist nature of the alliance; it was manifested above all in the announcement of the "Brezhnev doctrine."* Parallel with Soviet attempts to use the pact to improve Eastern European integration were efforts to use the Council for Mutual Economic Aid (COMECON).

The crisis in Czechoslovakia was an important experience for the non-Soviet allies: it gave them a dramatic lesson in the limits of diversity that Soviet leaders were willing to accept. Since 1968 the pact has continued to be used in a conflict-containment role, by word and sometimes deed. In the latter respect military movements have been used to threaten the independently minded Romanians and Poles. The allies have learned from each other's experiences that there is some room for maneuver within the pact if haste is avoided and certain ground rules are obeyed.

The constraints maintained by Soviet domination of the pact provided the setting within which the Eastern European countries became involved in the delicate maneuverings in the early 1970s associated with the lengthy prologue to the European security conference (CSCE). During this time the Warsaw Pact was unusually busy at the political level. This activity declined after Helsinki in 1975. However, there was one organizational development of note, the creation of the Council of Foreign Ministers in 1976, to advise the pact's Political Consultative Committee (PCC) on foreign policy. As with the Budapest reforms, this did not represent any decentralization of decision making; it was merely another expression of the Soviet attempt to tighten its grip on its allies. Following the decline of the pact's political activity in the years after CSCE, there was something of a revival in the early 1980s as the pact was used by the Soviet Union to secure support for its arms control proposals and for its "peace offensive" directed against NATO's proposed deployment of new theater nuclear missiles. Of particular importance in this respect was the January 1983 meeting of the PCC in Prague, when a "new grand peace proposal" was unveiled. It did not bring a positive response from the West, but its multilateral character helped to strengthen the Soviet initiative.

In the political sphere, but even more so in the military, a major feature of the Warsaw Pact has been its high degree of "integration"—in

---

*The "Brezhnev doctrine"—a designation used in the West but not in the Soviet Union—first appeared in an article in *Pravda* on 26 September 1968 under the title "Sovereignty and the International Obligations of Socialist Countries." The Brezhnev doctrine conceived the Socialist commonwealth as an organic unity, in which "every Communist Party is responsible not only to its own people but also to all the socialist countries and to the entire Communist movement." Stripped of its code words, it was a doctrine of limited sovereignty to legitimize Soviet interventions to defend the gains of (Moscow-defined) socialism. Operationally, in Eastern Europe, this means that the Communist parties in the region must maintain a monopoly of power and must not be completely independent of Moscow's wishes, while the governments of those countries should not leave the Warsaw Pact. One operational matter that is unresolved concerns the boundaries of the Socialist commonwealth: do they stretch beyond Eastern Europe?

marked contrast with NATO. However, this integration is of a very special kind: it has been, and for the foreseeable future will continue to be "integration" on Soviet terms. Although the pact is formally an alliance between independent sovereign states, it is quite clear who integrates whom. Operationally, the "Brezhnev doctrine," which limits sovereignty, is more important than the fine-sounding words about independence contained in the preamble to the Warsaw Treaty. The Soviet Union would like even more integration, and its ideas have even included a supranational force. But these have been opposed, notably by Romania.

One important manifestation of the integration achieved within the pact has been its high degree of standardization in equipment, doctrine, and military arrangements (for example, conscription obligations). The Soviet Union is the source of all major weapons and equipment, and much of the minor ones. The result is an impressive interchangeability of equipment between the national forces; this is in marked contrast to the arms bazaar that constitutes NATO's order of battle. In addition, the various national armies are organized on Soviet lines, with common tactical doctrine, training, and command-and-control arrangements. The advantages in operational efficiency should be obvious; as long, that is, as the Soviet command has got the right answers itself.

When it comes to financial burden sharing the picture of close integration breaks up somewhat, since economic difficulties and rising defense costs through the 1970s have intensified resource allocation issues in Eastern Europe. There always was a dichotomy between the northern and southern tier allies in terms of their readiness to meet the levels of defense spending required by the Soviet Union, but the Soviet problem became both more serious and complex at the end of 1978. After again demanding greater defense spending from its allies, the Soviet Union found that Romania would not dig in its pockets, while Poland could not. As with NATO, the problem of burden sharing is likely to grow rather than diminish.

In addition to providing some resistance to increased defense spending, the non-Soviet allies have also stood out—with varying strength—on the matter of Soviet suggestions to extend the pact's military coverage to the Far East or even Southeast Asia. None of the non-Soviet members has wished to be drawn by the Soviet Union into the latter's conflict with China, or into the problems with Vietnam. Although the room for maneuver for the Eastern European allies is limited, in these and other ways they show a degree of independence that does not neatly fit the cold war stereotypes of these countries, and of their relations with the Soviet Union. In 1983 and 1984 the Soviet Union had to face a small but significant amount of opposition to its plan to site new tactical nuclear missiles in East Germany and Czechoslovakia, ostensibly in response to the threatened Western INF deployment. The expressions of disagreement were met by firmness from Moscow, and in 1984 the missiles were deployed; this action was accompanied by a propaganda campaign in Eastern Europe, which stressed that it had been necessary in order to counter NATO's escalation of the arms race. During this period, Moscow also adopted an increasingly tough stance toward West Germany. This too provoked much disagreement between Moscow and various Eastern European regimes, since they had come to regard West Germany as more of a magnet than a bogey in the years following Ostpolitik. Moscow leaned on its allies, and at the end of 1984 the planned visits to Bonn of both the Bulgarian and East German leaders were called off. There were other signs of diversity. There were even reports that Romania might withdraw from the Warsaw Pact when it came up for renewal in 1985. Coinciding with a period of relative uncertainty at the top in Moscow—the two and a half years of funerals following Brezhnev's death—there were more public differences of opinion than at any time in the history of the Warsaw Pact (though they were not as immediately serious as, for example, the developments in Hungary in 1956 or Czechoslovakia in 1968).

In the first half of the 1980s, as on other occasions, the Soviet Union exerted its control over the region. Nevertheless, these signs of diversity were an important reminder to those Western observers whose minds have been frozen in the attitudes of the 1950s that Eastern Europe exists in the 1980s. One of the features of the latter is the importance and attraction of Western Europe to many in the eastern part of the continent, and the continuing and possibly growing differences in perspective between the Eastern European peoples and the leadership in Moscow. Looming Soviet power still dominates the region, and will for the foreseeable future, but it cannot prevent all pressure for change. In this sense the Warsaw Pact is bound to remain an alliance "in transition."[58] Significantly, despite the grumbling that had developed within the Soviet bloc over the previous couple of years, the leaders of the Eastern European regimes voted unanimously to renew the Warsaw Pact in April 1985, for up to thirty years (as in 1955). The ceremony was presided over by Gorbachev. Under its new and more energetic leader, it seemed that Soviet policy in Eastern Europe, as elsewhere, would remain essentially the same, but be discharged more effectively.

Despite all its efforts to strengthen the cohesion of the Warsaw Pact, by political and economic as well as military methods, the Soviet Union faces considerable problems in the region, its priority area in foreign and security policy. Diversity will remain, and possibly grow; criticism of the Soviet Union would increase. Economic and political problems will undoubtedly rumble on, and crises—big and small, single and simultaneous—cannot be ruled out. Preserving the status quo, and managing what change has to take place will remain a preoccupation in Soviet policy. Without doubt, the Warsaw Pact will continue to play a crucial role in Soviet strategy. The Soviet task will be helped of course by the self-restraint of the Eastern European countries themselves since they have become familiar with the rules of the game after 1956 and 1968. They know that if they step outside the bounds of Soviet tolerance, they can expect military pressure, and no help from anyone. From the record, they can have little doubt that the Soviet Union will use force if it thinks it necessary. Against this, however, must be set the fact that in three sets of crises (Czechoslovakia, Romania, and Poland) the Soviet Union showed patience and tried first to resolve matters by other than military means. The Soviet Leaders know that intervention is both costly and risky, and that it introduces an undesirable degree of unpredictability into international politics.

The arguments favoring restraint on the part of the Soviet Union must have been powerful during the Polish crisis of 1980–81, since, if force had been used by the Russians, it would not have come as a surprise, given the extremely serious nature of the crisis. In Poland, the national Communist party had lost control of affairs. In consequence, as *The Economist* later put it, "The withering away of the communist party, under the glare of the Soviet empire's first genuine worker's movement, has put a question-mark over the empire's whole ideology."[59] If unchecked, this development would have posed a serious challenge to Soviet domination of Eastern Europe and, over the longer term, even to the authority of the Communist Party of the Soviet Union (CPSU) itself in Moscow.[60] The leaders in the Kremlin, like their counterparts in the White House, worry about dominoes toppling in their direction. On at least two occasions during the Polish crisis it appeared that Soviet military intervention would occur. Nevertheless, the mere threat of force did not stop some changes taking place in Poland.[61]

The problem and danger for the Russians in Poland still simmers, though temporarily a cap was put on it by the imposition of repressive martial law in December 1981. It is not known whether General Jaruzelski's military intervention in his country's own affairs was the result of direct Soviet pressure or whether it was an independent initiative in response to the growing threat of Soviet action. In either case, the internal crackdown represented the best of a bad job from the Soviet perspective. Once again, looming military power had exercised an authority over men's minds.

In Eastern Europe, as elsewhere, the Soviet Union equates its security with military strength. Given the growth of Soviet military power since the 1960s, therefore, this should mean that Soviet confidence about its position in Eastern Europe has increased. But such has not been the case; problems have grown, not a sense of security. Largely because of this unease the Warsaw Pact will continue to have a major role in Soviet efforts to maintain control over the region. The continuance and success of the pact is too important for the Soviet Union to end it, even though the Soviet military position in the region is also buttressed by a series of bilateral treaties. Nevertheless, it seems unavoidable that the Eastern European states will continue developing in different ways and at different rates, and that changes in one country will have implications elsewhere. This means that the region will remain a significant focus of attention for students of international politics, as well as of the Soviet Union, for many years to come.

Although the Soviet leaders must regard their allies with varying degrees of suspicion, and although the pact is under some strain as a result of a variety of political, economic, and diplomatic changes, Western observers should hesitate before making any incautious predictions about the future of the region. There are, to be sure, divergences of interest between the Soviet Union and its allies, but the Socialist commonwealth in Eastern Europe is held together by more than Soviet military might. The party establishments, for example, have close relations with, and in some cases are dependent upon Moscow, while the national military establishments of the allies have usually been quite loyal to the neighborhood superpower. In addition, there is the NATO (and especially the German) threat, which pushes these countries under Soviet protection. From several points of view, therefore, the "security" of the regimes in Eastern Europe cannot be decoupled from Soviet power. As well as offering the regimes of the non-Soviet allies some security, the pact also provides an institutional framework for their relationship with the dominant and domineering regional power and gives them some legitimacy and diplomatic clout. For its part, the Soviet Union sees the pact as militarily and politically vital to ensure Moscow-defined Socialist solidarity in Eastern Europe; it does this by denying the allies complete control over their own forces and by maintaining the infrastructure for Soviet intervention. Through the pact the Soviet Union maintains a military and political buffer against the West and provides itself with greater military potential in the event of war. As the chief multilateral organization for deterring NATO and for managing change in Eastern Europe in a Soviet-desired direction, the Warsaw Pact has little reason to fear redundancy.

## *Sino-Soviet Military Relations*

As if the problems with comrades on the western frontier were not bad enough, Soviet military relations with Communist China deteriorated steadily through the 1960s, and by 1969 had reached the brink of war. The image of a monolithic Communist threat that had so exercised many Western minds through the 1950s finally shattered, and after 1969 the prospect of a Sino-Soviet rapprochement seemed to some observers as impossible as had the prospect of a Sino-Soviet conflict a decade earlier. In any event, the split was one of the decisive events in postwar international history, and a major blow to Soviet hopes. By the end of the 1970s, with the Polish crisis on one side and Chinese hostility on the other, Westerners were able to quip that Brezhnev's Russia was the only country in the world surrounded by Communist enemies.

In February 1950 the Soviet Union and the People's Republic of China signed the Treaty of Friendship, Alliance, and Mutual Assistance.[62] This Soviet-dominated alliance had confronted the Western powers with two apparently united Communist giants, and in so doing had increased their individual deterrent power and general political strength, for example, during the Korean War.[63] From the late 1950s onward, however, the strains in the Sino-Soviet military relationship grew, and both the political and military relationship deteriorated rapidly after 1960. While military relations with the United States became more manageable through the 1960s from a Soviet viewpoint, those with China appeared increasingly dangerous, judging by both word and deed. In 1969 there were armed clashes and talk of war. But war did not break out; nor, subsequently, did the normalization of relations sought by the Soviet Union. The Chinese terms were too high. The Sino-Soviet Treaty of Friendship nevertheless lingered on until April 1979, when the Chinese government finally announced that it was no longer relevant; it was allowed to lapse in February 1980.

Relations had never really been cordial between the two Communist superpowers, even during the negotiation and life of their alliance. In particular, the Chinese resented Soviet high-handedness and unhelpfulness. Hindsight reveals that the clash of interest and perspective between the two countries was bound to have produced tension and acrimony at some time or another. Sino-Soviet relations were burdened by a legacy of mistrust, a history of personal rivalries and difficulties between their leaders, territorial conflicts, ideological differences, clashes of opinion about international developments, and racial distaste. Coexistence was always uneasy between the two giants, the one an ambitious but insecure superpower, the other a technologically underdeveloped country whose population constituted a quarter of humanity. Their common six-thousand-kilometer border, some of it disputed, was a likely place for trouble.

In March 1969 serious clashes took place in a disputed area in the Amur-Ussuri region. Initially, Chinese soldiers moved on Damansky Island, one of several pieces of territory claimed by China. Soviet frontier guards were killed. Two weeks later Soviet forces carried out a strong counterattack that took them into Chinese territory. Estimates put Chinese deaths at about eight hundred and Soviet ones at about fifty. Although tension remained high in subsequent months, the fighting continued only in words. Negotiations took place on the border problem, then and in subsequent years, but the talks have been broken off as frequently as they have been begun.

According to one observer[64] China had been identified as the Soviet Union's major threat even before the border clashes. Harrison Salisbury traces this perception to 1966, when a nuclear China, in the grip of the Cultural Revolution, went out of its way to demonstrate its hostility toward the Soviet Union. After the clashes of 1969 there could be no doubt about Soviet priorities. The United States might remain the "ultimate" enemy in terms of potential, but China was clearly the country with which the Soviet Union was most likely to fight in the foreseeable future. A concerted and impressive Soviet military buildup in the Far East ensued.* This included the construction of the new BAM railway, to the north of the vulnerable Trans-Siberian line. The Soviet buildup indicated a determination not to surrender any territory and to ensure that any negotiations that occurred between the two countries took place against a background of evident Soviet military potential. If any new clashes were to break out, the Soviet Union was determined to win them.

Despite the Soviet military buildup in the Far East, the Soviet leadership has not proved able to translate destructive capacity into diplomatic influence in its dealings with the Chinese. In the immediate aftermath of the 1969 border clashes there was evidence that the Soviet leaders contemplated a surgical strike against China's growing military potential.[65] As time has passed, so have whatever attractions this option may have had. And other options have also been narrowing, since the Chinese can now meet Soviet threats with some sort of counter at all levels—from the challenge of "people's war" against a Soviet threat of general invasion to the posing of a minimum deterrent against Soviet nuclear threats.[66] The Chinese have been modernizing their military machine, albeit slowly, and they are now more confident of their deterrent capabilities. The potential costs for the Russians of using brute military force against the Chinese are therefore growing every year, and as they do so, the diplomatic utility of Soviet military power will diminish.

Whereas the interpretation of Sino-Soviet relations that has just been given looks toward a period of uneasy standoff between the two countries, some observers are less sanguine. One strategic writer, Edward Luttwak, has argued that since U.S. strength rules out the prospect of Soviet aggression against that power, and since Soviet aims in Western

*Soviet manpower on the border rose from 15 divisions in 1968 to 45 by 1974. There are now 52 divisions; 35 of those are said to be in Category 1 or 2 readiness. Nearly half the Soviet total of SS-20s (108) are deployed in the Far East. *The Military Balance 1983–84* (London: International Institute for Strategic Studies, 1983), 14–16.

Europe can be substantially achieved without direct attack, then China is destined to be the main target in the years ahead for what he sees as the Soviet empire's pent-up military energies. Luttwak has argued that the Soviet Union now possesses some "realistic war options" against its main enemy, Peking, and that the strategic security of the Soviet Union could be enhanced by "a limited but map-changing war upon China." He sees the two countries as "almost mechanically preordained to hostility towards each other."[67]

While such warnings as those of Luttwak are useful reminders of the dangerous possibilities in the situation, most observers believe that since both countries would suffer horribly in any conflict, reason demands coexistence. This suggests that Sino-Soviet relations will continue in the immediate future, as in the immediate past, to fluctuate between cold war and incipient détente.

In the early 1980s the Soviet Union occasionally came forward with offers of improving relations with China, but they got nowhere, mainly because the Chinese required concrete expressions of accommodation that the Soviet leaders did not feel able to offer; the terms proposed by the Chinese demanded too great a change in existing Soviet policy. The Chinese have had three specific demands. First, they have looked toward a Soviet withdrawal from Afghanistan (the Chinese were very hostile to the Soviet intervention and have been giving aid and support to the "rebels"). Second, they want the Soviet Union to persuade its ally, Vietnam, to withdraw from Kampuchea (China and Vietnam fought in February 1979, and the Soviet Union continues to back Vietnam). Third, the Chinese want a reduction of Soviet military power massed on their common border. The Soviet leaders have not been able to accept such terms from a country with which it has so many clashes of interest. There is the inherited hostility between them; unfavorable demographic trends; Chinese claims on Soviet (once Chinese) territory; Soviet fears that the Chinese might take advantage of any conflicts between the West and the Soviet Union; Soviet anxieties about the normalizing of relations between China and Japan (and the prospect of these two Far Eastern giants growing closer to the United States and Western Europe); and Soviet concern about the problems Chinese interference might cause to Soviet interests in southern and Southeast Asia. In sum, China will represent in the years ahead an important element in the hostile external environment that will confront Soviet foreign and defense policy planners.

The Soviet Union has military options for dealing with the problem of China, but every year they become costlier. This should increase the desire of Soviet leaders to relax tensions with China; but Soviet leaders will certainly resist détente if it can only be achieved on terms demanded by the Chinese. If relations are to improve, therefore, circumstances must change so as to allow the Chinese to drop their demands, or at least greatly reduce them. Such an outcome is possible, since the main issues could become less important to the Chinese. However, for the immediate future a Sino-Soviet rapprochement now seems almost as unthinkable as did a Sino-Soviet split in the 1950s. As that split showed, however, international politics are prone to surprise. But for the time being, while an uneasy standoff persists between the Socialist giants, the West profits.

## *Deterrence and Arms Control*

The interplay between Soviet and Russian thoughtways and the evolution of the internal and external environment has produced, over time, a characteristic set of *Soviet* attitudes to defense—what some have called a distinctive "strategic culture"—to underline that the continuity of attitudes involved is more than mere "policy."[68] The dominant features of this Soviet strategic culture include a highly security-conscious leadership, a society in which military values and burdens are accepted without hesitation, a great suspicion of outsiders, an understanding of the dangers of being weak, a tendency to see the struggle of international politics in crude power terms, a habit of military overinsurance, a belief that peace and stability are best secured through one's own strength, and a predisposition on the part of decision makers to prefer the traditional approaches of the military over the untested theories of civilians.

As a result of this strategic culture, Soviet defense planners have come to conceive strategic concepts in a rather different way from their Western counterparts. It would be odd if this were not so, though some Western observers have given insufficient credit to the peculiar context out of which Soviet military and foreign policies have grown. When looked at from the Soviet perspective, much of what is often labeled aggressive, unsophisticated, or adventurous by Western observers appears in a rather different light.

As was mentioned at the beginning of this chapter, the unique Soviet-Russian experience has given different meanings to the major strategic concepts. Most importantly, for example, Soviet military thinkers have not conceived "deterrence" in the same way as those mainly American nuclear theorists of the 1950s and 1960s who have so dominated Western thinking. A great deal has flowed from this, in terms of the way we interpret particular aspects of Soviet military behavior.

At the outset, Soviet strategic thinking has not made the same doctrinal distinction between deterrence and defense that Western strategists have drawn.[69] In Soviet thinking the one is an extension of the other, or even synonymous. The Soviet conception is the traditional military view, which comprehends deterrence in terms of a threat by an impressive war-fighting capability. The Anglo-American conception, on the other hand, crystallized through the 1950s and 1960s into the idea of mutual deterrence; the capabilities and doctrines necessary for this strategy were significantly different from those desirable for an effective war-fighting capability. Furthermore, the concept of mutual deterrence includes the idea of being solicitous about the survivability of the adversary's retaliatory forces, as well as one's own, and of being willing to maintain one's own vulnerability, as well as working to ensure that of the adversary. Such approaches are alien to the traditional viewpoint, and hence to Soviet strategic culture.

Because Soviet military thinkers have seen deterrence and defense to be synonymous, they have given much more explicit attention to fighting a nuclear war than Western (especially civilian) strategists. There has been no mental blockage caused by the thought that deterrence might fail. Consequently, Soviet planners have always attached a higher priority to the operational aspects of nuclear strategy, though in the late 1970s there was a marked change in U.S. thinking in this direction. This means that as well as seeking to ensure that they have an assured destruction capability, Soviet planners have also tried to develop damage limitation capabilities (including plans for preemption, air defense, and civil defense, for example)[70] and war-winning strategies (including rapid offensive action across Europe).[71] The adoption of a war-fighting strategy does not mean that the Soviet leaders want war or think it inevitable or intend to take advantage of the first "window of opportunity" they see. On the contrary: war-fighting is simply their theory of deterrence. It is the oldest approach to deterring enemies, and it is one that springs naturally from the traditionalist elements of the strategic culture and the military-dominated strategic policy-making process.

One consequence of this traditionalist approach to deterrence has been an old-fashioned preference for the idea of military superiority. In the American strategic community—particularly among those dubbed "defense intellectuals"—parity and balance became almost articles of faith in the late 1960s and early 1970s. In an era of overkill, superiority was seen as a pointless goal.[72] But the Soviet military leadership has never seen it in that way. It has preferred superiority; politically, it believes that numbers have an impact on one's adversary and on third-party perceptions of the strategic confrontation. Militarily it knows that numbers give options and depth. What is at issue in Soviet military thinking is not whether superiority is desirable in principle, but whether the nation can achieve it in practice. And because the leadership is doubtful about the latter its political masters have been more or less willing to accept rough parity with the United States. They would feel safer with superiority, but in practice they have learned from experience that if they race too hard, they will only provoke the United States, and that a provoked United States is not only dangerous, but will always win an arms race. So despite their preferences for overinsurance, Soviet leaders have eschewed superiority as a goal in the recent past. Instead, in practice, they have hoped to achieve "parity plus," that is, rough equality, but with margins of advantage when they can be achieved.

The advent of parity was one of the themes of Brezhnev's major pronouncements in the early 1970s, but he warned that the Soviet Union would respond to any U.S. attempts to obtain military superiority over the Soviet Union. When in the late 1970s increasing numbers of American observers argued that just the opposite was taking place, and that the Soviet Union was gaining a significant lead in the arms race, Brezhnev began to stress that the Soviet Union did not seek strategic superiority. This has been the position adopted by Soviet spokesmen since that time: denouncing the goal of superiority has become a settled part of Soviet policy.[73] In arms control this has meant that their objective has been parity or, if possible, "parity plus."

In the late 1960s Soviet military and political leaders entered the SALT process with fewer expectations about arms control than their Western counterparts.[74] Probably to their surprise, however, they did rather well out of SALT I, both politically and militarily. American progress in ABM technology was constrained, a satisfactory agreement on ICBMs was achieved, and, ten years after the Cuban missile crisis disaster, strategic parity with the United States was formalized. The Soviet leaders hoped for more of the same out of SALT II, which began at the end of 1972, but their hopes were disappointed. SALT II sought "balances" in more complex areas of the strategic confrontation. Agreement would have been difficult under most circumstances, but the prolonged negotiations happened to coincide with growing mistrust on both sides. Soviet worries about the Americans were almost a mirror image of American complaints about them during the same period: they were suspicious of the U.S. arms buildup, they doubted the depth of the U.S. commitment to détente, and they were mistrustful of American positions on arms control. Nevertheless, the Soviet leaders remained committed to SALT II, not least because the emerging agreement seemed to be offering them some advantages. They were therefore disappointed in 1979 when the Americans—who were increasingly mistrustful of Soviet behavior and increasingly worried about developing Soviet advantages—pulled out of the process. SALT II had dragged on far too long for its own good.

After the collapse of SALT II, arms control fell into disarray. Nevertheless, the Soviet leaders tried for a short time to use the INF talks in Geneva to further their aim of halting the arrival in Western Europe of cruise missiles and Pershing missiles from the United States, weapons the Russians claimed were escalating the arms race in the European theater. When those weapons began to be deployed at the end of 1983, the Russians carried out their threat and walked out of the INF talks, together with the other major negotiations in which they were involved, notably START and MBFR.* A fundamental gulf existed between the two sides: it was not simply the result of one being *for* arms control and the other *against*, or the result of one being *evil* and the other being *good*, but rather it was the outcome of a clash of interests and perceptions about the issues involved, which in turn arose from their different geographies, strategies, technologies, experiences, and concepts of stability. As a result, even when they agreed about the numbers of weapons involved, they had what they considered legitimate grounds for disagreeing over what it was proper or sensible to count in a particular "balance."

Arms control went into hibernation in 1984, though the process was resumed again at the beginning of 1985. The superpowers restarted arms negotiations when the Soviet Union, having broken off the talks for tactical reasons, decided that its ploy had failed: it had failed to turn Western opinion against the continued deployment of cruise and Pershing II, and it had led to adverse propaganda about Soviet attitudes to arms control. When Reagan's reelection appeared certain, the Soviet Union indicated its wish to return to the negotiating table, and talks began in Geneva in March 1985. There were no changes in the basic positions of the two sides while the talks were in abeyance. From the outset, then, the talks were accompanied by pessimistic expectations; all the old mistrust remained, while a number of new complicating factors had arisen. There was more talk of arms control cheating than of likely success. From the Soviet side there remained the suspicion that U.S. arms control proposals were designed primarily to freeze U.S. advantages, legitimize the building of new U.S. systems, consolidate existing U.S. doctrinal preferences, and place the Soviet Union at a disadvantage in propaganda and strategic terms. Among the new complicating factors were the actual deployment of U.S. INF in Western Europe, the Soviet forward deployment of new missiles in Eastern Europe, and, above all, the rise of the Star Wars issue—something the Reagan administration declared to be "non-negotiable," but which the Soviet Union committed itself to stop. As time goes on, there will be further complications arising out of the deployment of new weaponry: in this case U.S. ship-launched cruise missiles and mobile ICBMs by the Soviet Union.

Although pessimism was the prevailing mood at the resumption of the talks, there were some positive factors. Both superpowers had some—if not overpowering economic incentives for slowing down their arms competition. There were signs that groups in both countries believed that the "new cold war" had gone too far. The personality factor was more hopeful: it was widely mooted, at least during the election year of 1984, that Ronald Reagan wished to go down in history as a "peace-

*For a fuller account, see Ken Booth, "Disarmament and Arms Control," chapter 6 in the companion volume to this one.

maker," while Mikhail Gorbachev was expected to try to improve the Soviet image in arms control following the Soviet "walkout" at the end of 1983. The concrete indicator of a more constructive attitude on the part of the superpowers was the merging of the three sets of talks—on space weapons, strategic offensive forces, and INF—under one umbrella. On the Soviet side Gorbachev undertook various initiatives for propaganda reasons and perhaps to encourage more flexibility—he offered an SS-20 freeze and a nuclear test ban—but these were immediately rejected by the White House as meaningless gestures. Clearly, important differences of interest and outlook will continue to divide the superpowers on the question of strategic arms, with the Soviet Union stressing the importance of controlling ballistic missile defense in space, which the Reagan White House has declared to be non-negotiable, while the U.S. priority is curbing INF and strategic systems.

While eyes turned to Geneva at the start of 1985, the main issue in the short run was not whether deep cuts would result, but whether the superpowers still had enough commitment in the arms control process to abide by the unratified SALT II Treaty. With its provisions due to expire at the end of 1985, the building programs of both superpowers threatened its future. For the treaty limitations to survive—and their survival will be an indicator of the commitment of the superpowers to the mutual management of their arms competition—older systems will have to be retired on reconfigured. While not taking it as a sign of the Reagan administration's conversion to arms control, Western liberals were relieved in mid-1985 by the president's decision to retire an old Poseidon submarine in order to ensure that a new Trident system did not breach the treaty.

While this chapter has pointed to Soviet mistrust of U.S. arms control policy, it must be stressed that Soviet arms control delegations play the numbers game and the propaganda game, and they attempt all the national security–enhancing tactics of arms control gamesmanship. These facets of Soviet arms control behavior were evident in the first half of the 1980s, as earlier. In 1984, for example, Soviet spokesmen favored fine-sounding declarations about the no first use of nuclear weapons and about nuclear weapons free zones; but they did not make the concrete proposals in other areas that would have encouraged Western governments to think that such Soviet proposals were anything other than propaganda. The last point was illustrated in the talks about a possible worldwide ban on chemical weapons, when the Soviet Union proved unwilling to accept the Western position on the verification of production facilities; this was something almost all Western observers regarded as an essential step. Similarly, when the United States probably surprised the Soviet Union by agreeing to the latter's proposal for negotiations on space weaponry (albeit on terms that differed from those proposed) the Soviet Union reacted by strengthening its own position in such a way as virtually to ensure that its own initiative would collapse. There is every reason to be cynical when considering the arms control gamesmanship of both superpowers.

While it does not come easy to most Western observers, an understanding of Soviet arms control behavior must include an appreciation of how Western and especially American arms control behavior appears in mistrustful Soviet eyes. From this perspective, rightly or wrongly, U.S. behavior has appeared as uncooperative as has Soviet behavior to American observers. In Soviet eyes the "zero option" of President Reagan—the proposal that neither side deploy *any* intermediate nuclear forces—was preposterous; while the public pronouncements of President Reagan and Prime Minister Thatcher on INF seemed like an attempt to either bully the Russians into unilateral disarmament or make them appear responsible for collapsing the talks. In Soviet eyes these tactics, in addition to being crude, suggested that the Reagan administration was not interested in "serious" arms control negotiations but rather in propaganda or in agreements on its own terms.

Not without justification, both superpowers mistrust each other on questions of disarmament and arms control. This being so, it is not surprising that they have both accused the other of "cheating" in arms control agreements. Both have something of a case. In the mid-1980s the main focus of attention was the large phased-array radar built by the Soviet Union at Krasnoyarsk, which the White House has claimed to be suitable for ABM use and is therefore a violation of the ABM Treaty. The Soviet Union had defined the system as being for space tracking. The issue is more open than the Reagan administration has claimed, especially as there is a record of the White House rushing to judgment on such matters. But the development also illustrates the Soviet habit of pushing the treaties to the limit and exploiting whatever loopholes exist, or the nonvigilance of cosignatories.[75]

The Soviet leadership, like their counterparts in the United States, had become very cynical about arms control by the early 1980s. And the Russians always had less faith in the idea of the mutual management of the strategic confrontation through negotiations with their adversary. But despite their experience in the late 1970s and early 1980s and their pessimism about the immediate future, the Russians have not thrown arms control out the window. At a time when their relations with the United States have become uncomfortable in the extreme, they would value talks as a means of signifying their desire to maintain a minimally satisfactory relationship. In addition, talks might offer possible spin-offs in terms of propaganda (especially toward Western European peace movements). Negotiations might help slow down the drain of hard-earned cash into the defense sector and they might even constrain the development of new U.S. systems. The Soviet Union will explore arms control possibilities and will occasionally negotiate seriously and in a spirit of give-and-take (by their terms). But they will not accept a *diktat* or what they regard as an unsatisfactory deal. In this respect their walkout from the INF talks was entirely predictable. Equally, given the Soviet interest in restricting the proliferation of nuclear weapons, it was not entirely surprising that the Soviet Union signed an agreement with the International Atomic Energy Authority to open some of its civil nuclear facilities for inspection or that they were willing to discuss proliferation problems with the United States—despite all their differences of opinion—in the period before the 1985 review conference of the Non-Proliferation Treaty. Within the limitations set by mistrust and strategy, Soviet arms control policy under Gorbachev can be expected to show greater initiative and flexibility than in the recent past, but it will exhibit no less hardheadedness.

One consequence of the pervasive pessimism about arms control and the international situation generally during the first half of the 1980s will have been a decision on the part of the Soviet hierarchy to maintain its commitment to arms modernization. No less than in the past, they intend to try to ensure that they are not in a position of relative weakness. Soviet testing programs of new systems indicates the determination to stay with the United States in its strategic arms competition.

## *Current Problems*

Even though the major determinant of the Soviet arms buildup through the 1960s and 1970s was the challenge thrown down by its adversaries, the new military potential that was the outcome of the buildup has provided the leadership with strength and options which they did not formerly possess. Ironically, however, when and if they contemplate using this new power, their problems are at least as great as ever before.

In the event of war in Europe the Soviet leaders would be confronted by two enormous dilemmas. First, if war were to be contemplated by the Russians themselves, or even if it were forced on them, they would have to decide whether they should attack (or counterattack) with conventional forces or nuclear weapons. Second, if nuclear fighting did begin, they would have to face the problem of whether and how to begin controlling it. Soviet thinking about such matters is difficult to describe; perhaps they do not know themselves what they would do if these nightmare dilemmas suddenly became a hellish reality.

If the Soviet leaders decided to go to war in Europe for any reason, there is considerable disagreement among Western specialists as to whether Soviet forces would attack with conventional weapons only, or whether they would get in a first nuclear strike (their support of a no-first-use doctrine notwithstanding). Western specialists do agree, however, that the likelihood of a cold Soviet decision to go to war in Europe is very low indeed. Whatever gains some future Soviet leader might see from attacking NATO would surely pale into insignificance when set against risks that are almost too frightening to contemplate in practice. Nevertheless, the possibility of a Soviet attack must be contemplated by the Soviet Union's adversaries, since the avoidance of crazy decisions can never be guaranteed in international politics.

Were the worst to happen and war to begin in central Europe, military logic might suggest that the Soviet forces would use nuclear weapons from the outset.[76] If they did not, they would surrender the nuclear initiative to NATO, and the governments of the latter have consistently remained committed to a first-use strategy in some circumstances. But if the Soviet forces did in fact strike first with nuclear weapons they would expect an almost inevitable nuclear response; thus might begin a war of unimaginable horror. Consequently, since Soviet strategists believe that wars should be about politics, and that weapons should be subservient to strategy, they might well be inclined to take the gamble to withhold nuclear weapons and to seek to carry off a successful conventional offensive. The aim of Soviet strategy, therefore, would be to organize as large a conventional attack as possible, which could reasonably hope to achieve surprise. These forces would seek a rapid victory before NATO could respond with nuclear weapons; they would hope that NATO would dither and rapidly collapse. The Soviet Union's own battlefield and theater nuclear forces would be held back, to be launched at the moment they believed NATO was about to resort to nuclear war.[77] They might or might not plan to use chemical weapons. The need for surprise in such a scenario necessarily places some reliance on the forces of the "northern tier" allies, as was discussed earlier. But given likely uncertainties about the latter's performance, this is an additional constraint against Soviet military adventurism against Western Europe.

Soviet military planners, like their NATO counterparts, face many problems and ambiguities when it comes to devising an operational strategy in a nuclear environment. Neither side can escape the doctrinal compromises that have to be made. In the Soviet case this is well illustrated by their position on the second of the two major dilemmas they face when contemplating war in Europe, that of "limited nuclear options."

Soviet spokesmen have consistently said that introducing "rules of the game" into nuclear war is an illusion. Their declaratory position has been to assert that once a war becomes nuclear, it will escalate. "Control" is unlikely to succeed. Soviet pessimism in this regard, which is widely shared in Western defense circles, appears to be genuine. Consequently, they have been worried since the mid-1970s about the American government's talk about "limited nuclear options": such a doctrine threatens to make it easier for the two sides to slip into a nuclear war, if ever their troops become engaged. Against the Soviet declaratory position, however, must be set Soviet insistence that military strategy should be subservient to political objectives. This being so, and survival presumably being the primary political objective of any Soviet leader, it would be logical for them in practice to try to do their best to fight a "controlled" nuclear war. If nuclear war did break out, therefore, they might well seek to limit it, despite their prewar declaratory position. But it is not likely that they would be confident that control would last. Those responsible for Soviet nuclear weapons would therefore have very itchy fingers; they would be waiting to fire their weapons preemptively, at the first hint that NATO was about to "go nuclear."[78] The Soviet High Command would hope to avoid having to take the nuclear decision, but if it believed that NATO was about to release its nuclear forces, it would probably have little hesitation in giving the fatalistic and fatal advice to get a Soviet nuclear blow in first.

If there is a potential gap between the Soviet Union's declaratory policy and action policy on limited nuclear options, the same is not true on the vexed question of theater nuclear forces.[79] In the Soviet view, any weapon capable of hitting Soviet territory is defined as strategic, and therefore something to be taken very seriously. This outlook has been at the root of their position—and Western problems—in the INF talks. From the Soviet perspective, NATO's 1979 decision to deploy long-range cruise missiles and Pershing IIs in Western Europe represented a threatened escalation of the arms race, and more specifically an escalation of U.S. forward-based systems deployed in Europe. The Soviet Union claimed that these weapons threatened to upset the theater nuclear balance, and their claim was somewhat tenable; there are a number of such positions that can be honestly held on this complex issue.* Soviet defense planners have always placed great emphasis on theater nuclear forces, and they have created a large INF capability to cover both Europe and the Far East. In the late 1950s and early 1960s the Soviet Union deployed rather inaccurate and vulnerable SS-4s and SS-5s to counter U.S. forward-based systems in Europe. When these weapons were becoming seriously obsolete, the Russians started replacing them after 1977 with the SS-20, a more accurate, flexible, and mobile system, and one altogether better suited to the needs of a possible war in Europe. The Russians argued that this step represented a modernization rather than escalation. Unfortunately, the problem of continuous weapons innovation is such that one man's modernization looks very much like another's escalation.

From NATO's point of view, long-range cruise missiles and Pershing IIs—and the accuracy and short flight time of the latter makes them particularly worrisome for the Russians—were a response to the deployment and steady buildup of Soviet SS-20s. If the SS-20s were not bal-

*See chapter 6 in the companion volume to this one.

anced by comparable forces in the West, it was argued that the Russians would be left with escalation dominance in that band of capability. Soon after NATO had begun the deployment of its new missiles the Soviet Union introduced new ones of its own into Czechoslovakia and East Germany, as was mentioned earlier. These new missiles, added to the SS-20s and others already deployed, have been of growing concern in NATO circles because of the perceived vulnerability of NATO's military infrastructure. NATO's active and passive defenses against this threat are believed to need considerable upgrading; among other things, the enhanced Soviet capability has raised the issue of antitactical ballistic missiles (ATBM). And so it goes. Each side blames the other for escalating theater nuclear forces in Europe. It is an apparently irreversible spiral.

But nowhere in the mid-1980s have the dynamics of the arms race spiral been of more concern than in the prospective militarization of space. Large numbers of commentators in many different countries believe that if restraints are not quickly imposed on arms competition in space, then this will become yet another arena for costly, dangerous, and ultimately futile arms racing between the superpowers. In the aftermath of President Reagan's Star Wars speech of March 1983, Soviet political leaders and defense planners felt pressured and uncertain by U.S. ambitions and potential in space. But one thing at least seemed sure in the era of strategic speculation opened up by the Reagan administration's commitment to ballistic missile defense in space. That is, depite the president's claims, nuclear weapons would not be abolished as a result of defensive developments. The age of deterrence would not be superseded. That the Soviet Union would cease to develop its strategic offensive forces in parallel with the developing potential of a U.S. defensive screen was unthinkable. If there are no agreed restraints on space weaponry, then the Soviet Union, as ever, will respond in its own way to its familiar predicament of being outflanked by U.S. military technology.

Soviet defense planners have traditionally favored a mixture of offensive and defensive systems, and research into ballistic missile defense (BMD) continued through the 1970s.[80] Soviet progress in the development of ground-based ABMs has been such, it is believed in some quarters, that rapid expansion would be possible if the 1972 ABM Treaty were to collapse. However, it is also strongly believed by the U.S. Department of Defense, among others, that the United States has a significant lead in the technologies required for ABM systems, and even more so for space-based BMD.[81] This also appears to be the Soviet assessment, and so the stopping of President Reagan's Strategic Defense Initiative (SDI) immediately became a top priority for the Soviet leadership, though the Reagan administration insisted that the SDI was a non-negotiable and central aspect of its defense policy. Soviet concern about the SDI is based on Russian relative economic weakness, backwardness in the relevant technology, and fear that the SDI will form part of a U.S. first-strike capability. These concerns helped in 1984 to inject a new seriousness in Soviet thinking about arms control.

The seriousness with which the Soviet Union has looked upon U.S. technological potential in space weaponry was evident in a high-level report, published by a team of Soviet space scientists and strategists, which was discussed in the West early in 1985.[82] The report also brought out the range of possible Soviet countermeasures and the destabilizing character of the new weaponry. On the question of countermeasures, five possibilities were discussed: the destruction of the U.S. space platforms (for example by "space mines" or missiles); encasing ICBMs to make them invulnerable to laser attack; masking ICBM launches (for example, by smokescreens); changing the pattern of ICBM launches to confuse the defensive systems; and the use of decoys to help overload the potential of the defensive forces. The report argued that the countermeasures could be set up more quickly than the SDI and would cost only 1 or 2 percent as much.

The destabilizing possibilities in BMD development discussed in the Soviet report parallel fears expressed in the West. These include the apprehension that the SDI might have a role as part of a first-strike strategy by the United States against the Soviet Union. Even if it proves to be short of a perfect defense against Soviet missiles, the SDI will remain a pressing worry. To Soviet planners a partially successful U.S. BMD system could be used to mop up a significant proportion of those Soviet missiles left to retaliate against the U.S. homeland following a U.S. first strike. If the United States has such a capability, it might be tempted to strike first in a crisis. Soviet fears in this regard have been magnified by the U.S. counterforce buildup under President Reagan, the U.S. shift to a war-fighting strategy, and the constant refusal of the White House to agree to a no-first-use declaration. Such fears have been dismissed as groundless by U.S. supporters of the SDI, though since most of the latter were alarmists about a U.S. "window of vulnerability" in the late 1970s, they should have recognized the strategic logic in the Soviet case.

Although in the mid-1980s it has been generally assumed that if there is no agreement to restrain Star Wars technology, then the Soviet Union is bound to follow where the United States leads, the report of the Soviet scientists suggests the opposite may be case. While the issue remains unresolved, the report (together with other evidence) seems to show a genuine Soviet concern about the SDI. The Soviet Union may try to avoid the economic burden of Star Wars by relying instead on cheaper countermeasures. But all the immediate signs have been that the Soviet Union would prefer to avoid even this outcome: a zero option in space weaponry would be preferable to a costly, destabilizing, and intense arms race in space. In its arms control policy after 1983 the Soviet Union gave top priority to space issues. The SDI made arms control more important for the Soviet Union, but far less easy to achieve.

In the mid-1980s superpower competition in space tended to distract commentators from what had been a primary focus of attention for about fifteen years, namely U.S.–Soviet rivalry at sea. Those who did review these developments, however, merely saw the familiar action-reaction phenomenon at work in a different arena. But whatever the determinants of Soviet warship building, as with other areas of military procurement, the most striking fact to many Western observers has been that the Soviet Union has built up a range of impressive capabilities. And in the case of naval power, because of the inherent flexibility of warships, Soviet leaders have a variety of new options opening up. If this developing strength is unmatched by countervailing Western forces, the Russians will be left with a powerful fleet with a worldwide capability. This could be employed as an important new instrument of policy. The question is: will they simply use the extra power to discharge existing tasks with more effectiveness, or will they use it as the backdrop for a more assertive policy?[83] Time alone will tell.

If the past is any guide (and it might not be) Soviet military behavior will be cautious when it comes to projecting military force beyond the borders secured in the course of defeating Hitler. The charge of Soviet adventurism, sometimes applied to the way the Soviet Union has projected power in the Third World, will not stand up. Certainly the Russians have been revisionist in the Third World, but what they have done they have generally done with restraint. An exception was the Khrushchev era, and particularly the miscalculation that led to the Cuban missile crisis; on this occasion the Soviet Union could have been criticized for adventurism. That period apart, Soviet military behavior outside the Soviet bloc has not involved any serious risks.[84] The way they chose to support national liberation movements in the 1970s, as discussed earlier, is a good illustration of this. In addition, the unhappy experience of Afghanistan since 1979—whatever the final outcome—cannot be an encouraging one when it comes to contemplating any possible future interventions. The potential costs are rising, and there is no evidence that Moscow's calculation of what is an acceptable cost is inflating in proportion.

The question of acceptable costs is at the bottom of all strategic calculations, and in this respect it is important to repeat that the Russians have experienced war and know its terrors better than most. This means that we should be very skeptical about those who have claimed that the Russians believe that they can fight and win a nuclear war; that they believe, in fact, that a nuclear war can be an acceptable instrument of policy.[85] Those holding such views argue that when Soviet spokesmen deny the possibility of victory in nuclear war they are simply engaging in propaganda. The weight of evidence suggests the opposite.

Through the 1950s, and with decreasing conviction into the 1960s, Soviet spokesmen said that if nuclear war did break out for any reason—and there is no evidence suggesting that they would be seeking such an outcome—the Soviet Union would win.[86] From about 1971, however, Soviet spokesmen have consistently taken the opposite position; they have accepted the catastrophic consequences of nuclear war. And since 1982 this position has been supported with an understanding that the Soviet Union would not be the first to use nuclear weapons. Most Soviet spokesmen today—scientists, political leaders, commentators, and military spokesmen—warn that there would no victors in nuclear war. There are still some differences of opinion, as there are in the United States; however, the military viewpoint seems to have moved significantly from ideological fundamentalism (Marxist and Clausewitzian) to the sensible view that meaningful victory is impossible in nuclear war.

There has therefore been a significant evolution in Soviet thinking about nuclear war. As in other areas of life, those who speak for the Soviet Union talk in an ideological way, but they recognize the important realities. In addition to appreciating the change that has taken place in Soviet statements on these matters, it is also necessary to understand the thinking that underlies the words. In this regard, unless one takes the hardly tenable position that everything said or written in the Soviet Union is for the purpose of "disinformation," it is difficult to disagree with the analysis of Soviet thinking put forward on this matter by Robert L. Arnett.[87] He has argued that when Soviet spokesmen have said that "war is a continuation of politics" even in the nuclear age, there are three major points that must be kept in mind. First, the spokesmen are giving their views on the *causes* of war; it is analysis rather than prescription that they are offering. Second, statements to the effect that "victory" is possible does not mean that they believe that it is, in any meaningful sense. There is a gap between declaratory and action policy. "Victory" is being threatened for deterrent purposes (as it has by the Reagan administration, though with different words); it is intoned to maintain morale (and there are parallels in U.S. tactical manuals); and it is proclaimed for ideological purposes (guardians of Marxist-Leninist dogma could not accept anything but final triumph as the end point of the historical process). Third, ritualistic proclamations abut victory—made only by low-level spokesmen—are contradicted by authoritative negative statements about nuclear war as a practical instrument of politics, and by estimates of the probable consequences of such a conflict. In publications written for internal consumption (as well as in major statements that some would merely dismiss as being for Western consumption) Soviet spokesmen have stated that nuclear war cannot serve as a practical instrument of policy; they talk about the dire consequences of such a war.

While rejecting nuclear war as an acceptable instrument of politics, the

Soviet Union has made serious and consistent preparations to fight one. This is because it believes that an effective fighting force is the best deterrent to a nuclear war and that there still remains a finite possibility that such a war will be forced upon it. Below the nuclear level, the military instrument has a variety of utilities for the Soviet Union. Stephen Kaplan has shown, for example, that between 1944 and 1979 the Soviet Union used its armed forces as a "political instrument" on 190 occasions.[88] But Kaplan's book also showed, as other studies have, that Moscow has generally employed its military power with deliberation and caution. We are not presently confronted by a marauding Soviet bear, but by a cautious superpower with enormous and growing military muscle. If it is allowed to employ its military power successfully in the years to come, it might be tempted to rely increasingly on force as an instrument of policy (remember Brzezinski's description of the Soviet Union as a "one-dimensional superpower"). This is obviously not an encouraging prospect for the many countries that border it, since tough military behavior might be provoked by either ambition or insecurity. Given that defensive actions tend to be imperatives, whereas international ambitions are voluntaristic, it is well to recall the old warning that there is nothing so dangerous as a Russian on the defensive.

## *Conclusion: Soviet Strengths and Weaknesses*

As the Soviet Union looked toward the second half of the 1980s under its new leader, Mikhail Gorbachev, it faced many serious problems. The objective conditions—including major constraints—that Gorbachev inherited seemed to rule out dramatic change. Continuity in policy seemed to be the order of the day, though it was assumed that there would be those changes of tone and marginal changes of substance that most new leaders introduce. While Gorbachev, about whom relatively little was known when he took control, might in time initiate major changes in policy, for the immediate future broad continuity in foreign and strategic policy seemed likely—though both might well be discharged more effectively—as priority is given to domestic affairs.

In Soviet foreign relations, as in domestic affairs, Gorbachev's inheritance was not a comfortable one. In the first half of the 1980s, the world looked a rather complex and hostile place for Soviet foreign and defense policy planners. Anti-Soviet feelings were widespread; its adversaries were relatively buoyant; its own internal problems were considerable; and there were major constraints on its influence-building potential almost everywhere. In particular, relations between the superpowers had gone through their worst period for twenty years, being characterized by hectoring, the cutting off of contacts, the breakdown of arms control, increased mistrust, heightened arms competition, and "gutter language." There was good reason why Brezhnev in his declining years, and then his three immediate successors, should have felt something of that sense of insecurity that for centuries has possessed those who have looked out on the world over the walls of the Kremlin.

Although in a physical sense the Soviet Union has never been stronger, its foreign policy in the first half of the 1980s faced major challenges in every direction. Relations with the United States, the world's biggest military and economic power, were worse than at any time since the Cuban missile crisis. Relations with China, the other Socialist superpower and a quarter of humanity to boot, were strained. Relations had been poor for a number of years with Japan, which had the world's most vibrant economy and was being pressed to translate it into more military power by the United States. Hopes of exploiting anti-American, anti-NATO, and antinuclear sentiments in Western Europe seemed to have diminished, while fears about the future of Eastern Europe remained constant. Gains in the Third World were minimal and tended to be temporary; the historic opportunity to establish major influence seemed to have been lost. And instability rather than influence characterized such areas of priority interest in the Third World as the Middle East and southwest Asia.

All these disappointing developments—made worse by the poor Soviet economic performance—have had implications for Soviet defense planners. At the same time the latter have also had to face up to problems of a more directly military nature, particularly as a result of the challenges thrown down by U.S. defense policy under the Reagan administration and the need to keep abreast (or ahead) of new generations of weaponry—defensive and offensive, tactical and strategic, nuclear and conventional, on earth and in space.

Clearly, the Soviet Union has made great strides since the late 1940s. It is a superpower with enormous military strength, and it can claim the right to be involved in the international politics of most regions in the world. But it is a rather muscle-bound superpower. Nevertheless, superpowers of any type are difficult to handle. They expect deference from those around them; they demand loyalty from their allies; they want the best weapons; and they claim the right to interfere in the politics of distant regions.

The neighbors and adversaries of the Soviet Union must respect its considerable strengths: the sheer amount of men and material it possesses, its consistently high commitment of effort to defense, the patriotism of the people and their willingness to endure hardship, the ruthlessness of the Soviet leadership when called upon, the growing technological ability of the state, and so on. These are traditional qualities associated with the country, but the Soviet military machine is not without its problems. These include some inadequate technology, economic constraints, morale problems in the services, shortcomings in military training, organizational weaknesses, and skilled manpower shortages. In sum, there is a gap between the widespread popular image of Soviet military strength in the West, and the probable reality on the ground, in the air, and at sea in the East. There is less to the impressive parades in Red Square and the huge exercises across the oceans and western Russia than meets the eye. Western governments, like those of the Soviet Union, have regularly indulged in inflating the threat posed by the adversary.[89]

The present Soviet leadership is relatively confident that it will not be attacked from outside. In face of Soviet strength, such an act would be the height of folly—as, historically speaking, attacking Russia has always been. The Soviet leadership is also relatively confident that it can keep control of affairs at home. But with respect to both internal and external affairs, Soviet decision makers are much more worried in the mid-1980s than they were in the mid-1970s. Confronting a hostile external environment and uneasy about the future, they will continue to seek stability through strength, and they will try to maintain what they have, if necessary by adopting an aggressive manner.

Many of the Soviet Union's present and recent problems in foreign affairs have been of their own making: their tough behavior provokes others to respond in kind. But an isolated, suspicious, and ethnocentric leadership cannot easily act otherwise. The Soviet leadership is less able than most to appreciate the interests and perspectives of others in the intractable security dilemma that confronts us all. Soviet insecurity, as well as Soviet ambition, represents a threat to the Soviet Union's neighbors and adversaries; and these will not be able to feel comfortable while faced by a superpower that is politically edgy but militarily powerful in a world where instability is rife.

## NOTES

1. For a sound introduction to contemporary Soviet strategy see, among others, David Holloway, *The Soviet Union and the Arms Race* (New Haven: Yale University Press, 1983); John Baylis and Gerald Segal, eds., *Soviet Strategy* (London: Croom Helm, 1981); and Derek Leebaert, ed., *Soviet Military Thinking* (London: Allen & Unwin, 1981).

2. See Michael MccGwire, *Soviet Naval Developments: Capability and Context* (New York: Preager, 1973), 1–5, 31–33.

3. See Ken Booth, *Strategy and Ethnocentrism* (London: Croom Helm, 1979).

4. See Raymond L. Garthoff, "On Estimating and Imputing Intentions," *International Security* 3, no. 3 (Winter 1978): 22–32. Garthoff wrote some of the earlier important works on Soviet strategy, especially *How Russia Makes War* (London: Allen & Unwin, 1954) and *Soviet Military Policy: A Historical Analysis* (London: Faber, 1966).

5. See, for example, Hedrick Smith, *The Russians* (London: Times Books, 1976), chapter 12.

6. This is J. Malcolm Mackintosh's definition of "strategic policy" from the Soviet viewpoint, in "Soviet Strategic Policy," *The World Today* 26, no. 7 (July 1970): 272.

7. Many observers have noted this tendency. See, for example, ibid., 270.

8. Note the arguments of Rebecca V. Strode and Colin S. Gray, "The Imperial Dimension of Soviet Military Power," *Problems of Communism* 30 (Nov.–Dec. 1981): 1–15.

9. See M. Garder, *A History of the Soviet Army* (London: Pall Mall Press, 1966).

10. For Soviet strategy under Stalin and Khrushchev see Thomas W. Wolfe, *Soviet Power and Europe 1945–1970* (Baltimore: Johns Hopkins University Press, 1970).

11. For a thorough analysis of the Soviet use of the military instrument for political purposes, see Stephen S. Kaplan, *Diplomacy of Power: Soviet Armed Forces as a Political Instrument* (Washington, D.C.: Brookings Institution, 1981). A briefer analysis is Ken Booth, *The Military Instrument in Soviet Foreign Policy 1971–1972* (London: Royal United Service Institute, 1974).

12. Herbert S. Dinerstein, *War and the Soviet Union* (New York: Praeger, 1962).

13. On Soviet thinking about war see Peter Vigor, *The Soviet View of War, Peace, and Neutrality* (London: Routledge and Kegan Paul, 1975); Thomas W. Wolfe, "Communist Outlook on War," in Vernon V. Aspaturian, ed., *Process and Power in Soviet Foreign Policy* (Boston: Little, Brown, 1971), 401–25; and Graham D. Vernon, ed., *Soviet Perceptions of War and Peace* (Washington, D.C.: National Defense University Press, 1981).

14. The zigzags of Khrushchev's policy are discussed by Michael MccGwire, "Soviet Military Doctrine: Contingency Planning and the Reality of World War," *Survival* 22, no. 3 (May–June 1980): 107–13.

15. The standard work on Soviet risk taking is Hannes Adomeit, *Soviet Risk-Taking and Crisis Behaviour: A Theoretical and Empirical Analysis* (London: Allen and Unwin, 1982). See also A. L. Horelick and M. R. Rush, *Strategic Power and Soviet Foreign Policy* (Chicago: Chicago University Press, 1966).

16. Most attention in the West was focused on the novel Soviet naval developments in the late 1960s. John Erickson was one of the few analysts stressing the developing potential of the air force; see his *Soviet Military Power* (London: Royal United Service Institute, 1971).

17. For an introduction to this contentious issue, see MccGwire, *Soviet Naval Developments;* Michael MccGwire, Ken Booth, and John McDonnell, *Soviet Naval Policy: Objectives and Constraints* (New York: Praeger, 1974); Michael MccGwire and John McDonnell, *Soviet Naval Influence: Domestic and Foreign Dimensions* (New York: Praeger, 1977); Bradford Dismukes and James H. McConnell, *Soviet Naval Diplomacy* (New York: Pergamon Press, 1979); Rogert A. Bathurst, *Understanding the Soviet Navy: A Handbook* (Newport, R.I.: Naval War College Press, 1979); S. S. Goshkov, *The Sea Power of the State* (USSR, in 1976; Annapolis: Naval Institute Press, 1979); Bryan Ranft and Geoffrey Till, *The Sea in Soviet Strategy* (London: Macmillan, 1983).

18. This argument is summarized by Michael MccGwire, "Soviet Naval Doctrine and Strategy," in Leebaert, *Soviet Military Thinking*, 148–50, 156–61; and Ranft and Till, *The Sea in Soviet Strategy*, 166–75.

19. Dismukes and McConnell, *Soviet Naval Diplomacy*, passim.

20. MccGwire, "Soviet Naval Doctrine and Strategy," 170–75.

21. Ranft and Till, *The Sea in Soviet Strategy*, 102–7; Lt. Cdr. Floyd D. Kennedy, "Soviet Doctrine and the Role of the Aircraft Carrier," *Naval War College Review* 32, no. 1 (Feb. 1979): 48–58; Charles C. Peterson, "Aircraft Carrier Development in Soviet Naval Theory," *Naval War College Review* 37, no. 1 (Jan.–Feb. 1984): 4–13.

22. See MccGwire, "Soviet Naval Doctrine and Strategy," and Ranft and Till, *The Sea in Soviet Strategy*, 66–80.

23. MccGwire, "Soviet Naval Doctrine and Strategy," 175–76.

24. Erickson, *Soviet Military Power*, 93–95.

25. Kaplan, *Diplomacy of Power*, passim.

26. Booth, *The Military Instrument*, 32–37.

27. Michael MccGwire, "The Overseas Role of a 'Soviet Military Presence,'" in MccGwire and McDonnell, *Soviet Naval Influence*, chapter 2.

28. For different perspectives see Holloway, *The Soviet Union and the Arms Race*, 95–98; Edward Luttwak, *The Grand Strategy of the Soviet Union* (London: Weidenfeld and Nicolson,

1983), 53–60, 81–86; Fred Halliday, *Threat from the East?* (Harmondsworth: Penguin Books, 1982); and Jiri Valenta, "From Prague to Kabul: The Soviet Style of Invasion," *International Security* 5 no. 2 (Fall 1980): 114–41.

29. This was the main consideration, according to Henry Trofimenko, a leading Kremlin specialist on East-West relations. See *London Times*, 19 July 1983.

30. A good example of the single-decision (blueprint strategy) interpretation is Avigdor Haselkorn, *The Evolution of Soviet Security Strategy, 1965–1975* (New York: Crane, Russak & Co., 1978).

31. See McGwire, "Soviet Military Doctrine."

32. For different interpretations, see Dismukes and McConnell, *Soviet Naval Diplomacy*, and Ranft and Till, *The Sea in Soviet Strategy*.

33. For an overview see Robin Higham and Jacob W. Kipp, *Soviet Aviation and Air Power: A Historical View* (Boulder, Col.: Westview Press, 1977).

34. Marshall D. Shulman, "U.S.–Soviet Relations and the Control of Nuclear Weapons," in Barry M. Blechman, ed., *Rethinking the U.S. Strategic Posture* (Cambridge, Mass.: Ballinger, 1982), 85.

35. See Jim Harrison and Pyare Shivpuri, *The Russian Threat: Its Myths and Realities* (London: Gateway Books, 1983), 295.

36. Note, for example, the comments of ex-Secretary of Defense Robert McNamara, printed in *The Listener*, 10 June 1982, and in *The Guardian*, 9 August 1982; and the views of General Bernard Rogers, Supreme Allied Commander, Europe, in *The Guardian*, 29 September 1982.

37. See John Erickson, "Soviet Military Policy, Priorities and Perspectives," *Round Table*, no. 256 (October 1974): 369–79.

38. The following comments are based on Malcolm Mackintosh, "The Soviet Military's Influence on Foreign Policy," in McGwire, et al., *Soviet Naval Objectives*, chapter 2.

39. Holloway, *The Soviet Union and the Arms Race*, 109–15; Harriet Fast Scott, "The Soviet Decision-Making Process for National Security Policy," in Robert L. Pfaltzgraff and Uri Ra'anan, eds., *National Security Policy: The Decision-Making Process* (Hamden, Conn.: Archon Books, 1984), 87–108.

40. Matthew P. Gallagher and Karl F. Spielmann, *Soviet Decision-Making for Defense: A Critique of U.S. Perspectives on the Arms Race* (New York: Praeger, 1972).

41. This is one of the central theses of Roman Kolkowicz, *The Soviet Military and the Communist Party* (Princeton, N.J.: Princeton University Press, 1967).

42. Mackintosh, "The Soviet Military's Influence," 25–29.

43. William E. Odom, "The Soviet Military and Foreign Policy," *Survival* 17 no. 6 (Nov.–Dec. 1975), 276–81.

44. On the politics of procurement see Holloway, *The Soviet Union and the Arms Race*, chapters 6–8; Andrew Cockburn, *The Threat: Inside the Soviet Military Machine* (London: Hutchinson, 1983), chapters 4, 5.

45. A useful brief note about Soviet defense expenditure is included each year in *The Military Balance* of the Internation Institute for Strategic Studies (IISS). See, e.g., *The Military Balance 1983–1984* (London: IISS, 1983), 13–14. Also see Franklyn D. Holzmann, "Soviet Military Spending: Assessing the Numbers Game," *International Security*, 6, no. 4 (Spring 1982): 78–101; D. F. Burton, "Estimating Soviet Defence Spending," *Problems of Communism* (March–April 1983): 85–93; Cockburn, *The Threat*, 79–81, 279–80 has some skeptical comments.

46. The London *Times*, 28 December 1983. See the important article by Richard Kaufman, "Causes of the Slowdown in Soviet Defense," *Soviet Economy* 1, no. 1 (Jan.–March 1985): 9–31.

47. Note "Can Andropov Control His Generals?" *The Economist*, 6 August 1983, 43–45.

48. This conclusion is suggested by Cockburn, *The Threat*, in the light of the inefficiency, corruption, and lack of initiative within the military machine. See also *The Guardian* 12 July 1985.

49. A comprehensive picture of the "nation in arms" is given by Harriet Fast Scott and William F. Scott, *The Armed Forces of the USSR* (Boulder, Colo.: Westview Press, 1979), part 3.

50. A good introduction to the politics of the region is Teresa Rakowska-Harmstone and Andrew Gyorgy, eds., *Communism in Eastern Europe* (Bloomington: Indiana University Press, 1981).

51. For the evolution of the pact see Malcolm Mackintosh, "The Evolution of the Warsaw Pact," *Adelphi Paper* no. 58 (London: International Institute for Strategic Studies, June 1969); and "The Warsaw Pact Today," *Survival* 16, no. 3 (May–June 1974): 122–26; R. A. Remington, *The Warsaw Pact: Case Studies in Communist Conflict Resolution* (Cambridge, Mass.: MIT Press, 1971); Lawrence Caldwell, "The Warsaw Pact: Directions of Change," *Problems of Communism* (September–October 1975): 1–19; Dale R. Herspring, "The Warsaw Pact at 25," *Problems of Communism* (September–October 1980): 1–15; Christopher D. Jones, *Soviet Influence in Eastern Europe: Political Autonomy and the Warsaw Pact* (New York: Praeger, 1981).

52. Remington, *The Warsaw Pact*, 8, 56–93.

53. Mackintosh, "The Warsaw Pact Today," 122–23.

54. Jones, *Soviet Influence in Eastern Europe*, 1.

55. John Erickson, "Soviet Combat Force on Continent Grows," *The London Times*, 19 February 1974.

56. A. Ross Johnson, Robert W. Dean, and Alexander Alexiev, *East European Military Establishments: The Warsaw Pact Northern Tier* (New York: Crane, Russak, & Co. 1982), 1–2.

57. See Johnson, et al., *East European Military Establishments*, passim.

58. See David Holloway and Jane Sharp, eds., *The Warsaw Pact: Alliance in Transition* (London: Macmillan, 1985).

59. *Economist*, 13 March 1982.

60. The best book on the crisis is Neal Ascherson, *The Polish August: The Self-Limiting Revolution* (Harmondsworth: Penguin Books, 1981).

61. Michael Tatu, "Intervention in Eastern Europe," in Kaplan, *Diplomacy of Power*, 262–63.

62. Useful background to the subject of this section is Raymond L. Garthoff, ed., *Sino-Soviet Military Relations* (London: Praeger, 1966).

63. See, for example, Malcolm Mackintosh, "The Soviet Attitude," in M. H. Halperin, ed., *Sino-Soviet Relations and Arms Control* (Cambridge, Mass.: MIT Press, 1967), 193–226.

64. Harrison E. Salisbury, *War Between Russia and China* (New York: Bantam Books, 1970), 105.

65. See, for example, Henry Kissinger, *White House Years* (Boston: Little, Brown & Co., 1979), 171–73, 183–86.

66. A brief but first-class analysis of this situation is Gerald Segal, "The Soviet 'Threat' at China's Gates," *Conflict Studies*, no. 143 (London: Institute for the Study of Conflict, 1983). The first Chinese ICBM was tested in May 1980, the first MIRV (possibly) in September 1981, and the first SLBM in October 1982. See also Gerald Segal, "Strategy and Ethnic Chic," *International Affairs* 60, no. 1 (Winter 1983–4): 15–30.

67. Luttwak, *The Grand Strategy of the Soviet Union*, 97–105.

68. See Jack Snyder, *The Soviet Strategic Culture: Implications for Limited Nuclear Options*, R-2154-AF (Santa Monica: RAND, September 1977).

69. On Soviet thinking about deterrence see, among others, Peter Vigor, "The Semantics of Deterrence and Defence," and Geoffrey Jukes, "The Military Approach to Deterrence and Defence," in MccGwire, et al., *Soviet Naval Policy*, chapters 25 and 26; Donald Hanson, "Is Soviet strategic Doctrine Superior?" *International Security* 7, no. 3 (Winter 1982–83): 61–83; Henry Trofimenko, *Changing Attitudes towards Deterrence, ACIS Working Paper* No. 25 (Los Angeles: Center for International and Strategic Affairs, UCLA, 1981); John Erickson, "The Soviet View of Deterrence," *Survival* 24, no. 6 (Nov.–Dec. 1982): 242–49; Snyder, *The Soviet Strategic Culture;* Gerald Segal, "Strategy and Ethnic Chic"; Benjamin S.

Lambeth, "How to Think about Soviet Military Doctrine," in John Baylis and Gerald Segal, eds., *Soviet Strategy* (London: Croom Helm, 1981), 105–23; Fritz W. Ermath, "Contrasts in American and Soviet Strategic Thought," in Leebaert, *Soviet Military Thinking*, 50–69; Roman Kolkowicz, "U.S. and Soviet Approaches to Military Strategy: Theory vs. Experience," *Ortbis* 25, no. 2 (Summer 1981): 307–29.

70. On these aspects of Soviet doctrine see, among others, Holloway, *The Soviet Union and the Arms Race;* R. A. Mason, "Military Strategy," in Edwina Moreton and Gerald Segal, eds., *Soviet Strategy toward Western Europe* (London: Allen & Unwin, 1984), 165–203; Fred M. Kaplan, "The Soviet Civil Defense Myth," *Bulletin of the Atomic Scientists* 34 (March 1978): 15–20; (April 1978): 41–51; David R. Jones, "Civil Defense," in David R. Jones, ed., *Soviet Armed Forces Review Annual*, vol. 2 (Gulf Breeze, Fla.: Academic International Press, 1978), 289–350; Leon Goure, *War Survival in Soviet Strategy: USSR Civil Defense* (Coral Gables, Fla.: University of Miami, Center for Advanced Studies, 1976); Michael J. Deane, *Strategic Defense in Soviet Strategy* (Coral Gables, Fla.: University of Miami, Center for Advanced Studies, 1980). The *Soviet Armed Forces Review Annual* is an invaluable survey of developments in various aspects of Soviet military life. It contains reviews of the services, special surveys on general themes, and discussions of doctrine and theory—as well as historical essays and factual material.

71. For a skeptical view of their prospects, see John J. Mersheimer, "Why the Soviets Can't Win Quickly in Central Europe," *International Security* 7, no. 1 (Summer 1982): 3–39. On the much-debated question of chemical warfare, see Charles J. Dick, "Chemical and Biological Warfare," in Jones, *Soviet Armed Forces Review Annual*, vol. 5, 214–23; Cockburn, *The Threat*, 219–21.

72. Although in the later 1970s Kissinger altered his views somewhat, in Moscow on 3 July 1974 he declared: "And one of the questions we have to ask ourselves as a country is what in the name of God is strategic superiority? What is the significance of it politically, militarily, operationally at these levels of numbers? What do you do with it?" The Soviet High Command's answer to such a question, according to John Erickson, would simply be "Use it." See John Erickson, "Soviet Military Policy: Priorities and Perceptions," *Round Table*, no. 256 (October 1974): 370.

73. See William G. Hyland, "The USSR and Nuclear War," in Blechman, *Rethinking the U.S. Strategic Posture*, 60–64.

74. For insights into Soviet behavior during SALT see John Newhouse, *Cold Dawn: The Story of SALT* (New York: Holt, Rinehart and Winston, 1973); Thomas W. Wolfe, *The SALT Experience* (Cambridge, Mass.: Ballinger, 1979); Strobe Talbott, *Endgame: The Inside Story of SALT II* (New York: Harper & Row, 1979); Gerard Smith, *Double-Talk: The Story of the First Strategic Arms Limitation Talks* (New York: Doubleday 1980); Samuel B. Payne, *The Soviet Union and SALT* (Cambridge, Mass.: MIT Press, 1980).

75. For the Reagan administration's complains about Soviet behavior, see U.S. Department of State, "Soviet Noncompliance with Arms Control Agreements." (Washington, D.C.: U.S. Department of State, Bureau of Public Affairs, Special Report no. 122, 1 February 1985). For a sample of alternative views see P. Pringle, "Russian Radar 'Not Illegal,' " *The Observer*, 24 March 1985; and Karl Pieragostini, "Soviet Cheating? Reagan's Rush to Judgment," *ADIU Report*, 6, no. 1 (Jan.–Feb. 1984): 1–5.

76. On this contentious issue, see Joseph D. Douglass, *A Soviet Selective Targeting Strategy toward Europe* (Arlington, Va.: System Planning Corp., June 1977); and *The Soviet Theater Nuclear Offensive, Studies in Communist Affairs*, vol. 1 (Washington, D.C.: U.S.G.P.O., 1976), and, with Amoretta M. Hoeber, *Soviet Strategy for Nuclear War* (Stanford: Hoover Institution, 1979); and *Conventional War and Escalation: The Soviet View* (New York: Crane, Russak Co. 1981); Colin S. Gray, *Strategic Studies and Public Policy: The American Experience* (Lexington: University of Kentucky Press, 1982), 102ff.

77. Ibid.

78. On "launch on warning," see P. J. Bracken, The Command and Control of Nuclear Forces (New Haven: Yale University Press), 42–44, 83–84, 97, 222–23, 232, 244–45.

79. See the works of Douglass in note 76; Holloway, *The Soviet Union and the Arms Race*, chapter 4; Stephen M. Meyer, "Soviet Theatre Nuclear Forces, Part I: Development of Doctrine and Objectives," *Adelphi Papers* no. 187 (London: International Institute for Strategic Studies, Winter 1983–84).

80. See Sayre Stevens, "The Soviet BMD Program," in Ashton B. Carter and David N. Schwartz, eds., *Ballistic Missile Defense* (Washington, D.C.: The Brookings Institution, 1984).

81. Sidney D. Drell, Philip J. Farley, and David Holloway, *The Reagan Strategic Defense Initiative: A Technical, Political, and Arms Control Assessment* (Stanford: Stanford University 1984), 21.

82. The following account of the report is based on Stephen Shenfield, " 'Star Wars'—A Soviet View," *Detente*, no. 3 (May 1985): 10–12.

83. MccGwire, "Soviet Naval Doctrine and Strategy," 175–76.

84. See Kaplan, *Diplomacy of Power;* Adomeit, *Soviet Risk-Taking and Crisis Behaviour;* Booth, *The Military Instrument in Soviet Foreign Policy 1917–1972;* Jan F. Triska, "The Soviet Union: Reckless or Cautious?" in Jan F. Triska and David D. Finley, *Soviet Foreign Policy* (New York: Macmillan, 1968), 310–49.

85. Standard references for this view are Foy D. Kohler and Mose L. Harvey, *The Role of Nuclear Forces in Current Soviet Strategy* (Miami: Center for Advanced International Studies, University of Miami, 1974); Richard Pipes, "Why the Soviet Union Thinks It Could Fight and Win a Nuclear War," *Commentary* 64 no. 1 (July 1977): 21–34; T. K. Jones and W. Scott Thompson, "Central War and Civil Defense," *Orbis* 22, no. 3 (Fall, 1978): 681–712.

86. The comments in this paragraph are based on Stephen Shenfield, "Soviet Thinking about the Unthinkable," *Bulletin of the Atomic Scientists* 41, no. 2 (Feb. 1985): 23–25.

87. Robert L. Arnett, "Soviet Attitudes towards Nuclear War: Do They Really Think They Can Win?" in Baylis and Segal, *Soviet Strategy*, 55–74; and "The Consequences of World War III: The Soviet Perspective," in *The Soviet Armed Forces Review Annual*, vol. 5, 256–65.

88. Kaplan, *Diplomacy of Power.*

89. Cockburn, *The Threat*, especially 273–84. On some of the problems of assessing Soviet military power see Geoffrey Jukes, "Western Assessments of Soviet Strength," in Philip Towle, *Estimating Foreign Military Strength* (London: Croom Helm, 1982), chapter 3. Typically, official statements in the West paint a picture of consistent and growing Soviet conventional superiority in Europe. A number of sources in recent years point to rather different conclusions. See, for example, J. J. Mearsheimer, *Conventional Deterrence* (Ithaca, N.Y.: Cornell University Press, 1983), 165–88; O. P. Chaney, "The Soviet Threat to Europe: Prospects for the 1980s," *Parameters* (Journal of the U.S. Army War College), 13, no. 3 (1983): 2–22; B. R. Posen, "Measuring the European Conventional Balance: Coping with Complexity in Threat Assessment," *International Security* 9, no. 3 (Winter 1984): 47–88; M. A. Evangelista, "Stalin's Postwar Army Reappraised," *International Security* 7, no. 3 (Winter 1983): 110–38; and M. Urban, "Red Flag Over Germany," *Armed Forces* 4 (Feb., March, April 1985): 69–74, 110–15, 149–55.

4

# Chinese Defense Policy

JOHN BAYLIS

The task of studying the defense policy of any state is inevitably a difficult one, given the universal preoccupation with secrecy. For the Western student of Chinese defense policy the problems are particularly complex. Not only does he have to face the linguistic and semantic difficulties, but also those of a heritage, culture, and ideology so alien to his own. These problems are further compounded by the scarcity of reliable information available due to the lack of documentary sources, memoir material, and scholarly Chinese writings on the subject. The evidence available to the analyst therefore is, of necessity, fragmentary and often ambiguous. As Coral Bell quite rightly points out, "the observer is in no position to offer certainties, only a tentative assessment subject to the modification as further evidence comes in."[1] Despite these difficulties, one must attempt to interpret the defense policy of a country as important as China, especially as it emerges from its isolation to play a greater role in international affairs.

## *The Historical Legacy*

Throughout its long and impressive history China has always regarded itself as the center of world civilization. Even the term *China* (Chung-Kuo) meaning the "middle or central kingdom" conveys the idea of a "large universe revolving around a primary directing force" represented by China.[2] For much of its long history, Chinese military, cultural, and economic prosperity over its Asian neighbors enabled successive Chinese dynasties to establish China at the center of a system of tributary states, which were expected to be submissive and pay tribute to "the middle kingdom."

This feeling of China's superiority, however, received a severe setback in the nineteenth century when Chinese weakness and the advanced technology of many Western states enabled them to force the Chinese to open up their ports to foreign trade and cede their territory in a series of "unequal treaties." To a people as proud of their past glory and civilization as the Chinese, such domination by foreigners, and therefore bar-

barians by definition, was a humiliation that has provided a stimulus to all Chinese leaders since. This humiliation was summed up by the founding father of republican China, Sun Yat-sen, when he said, "We are the poorest and weakest country in the world, occupying the lowest position in world affairs; people of other countries are the carving knife and we are the fish and meat."[3]

The obsessive concern with their humiliations at the hands of Western states in the nineteenth century led to a debate that continued for much of the late nineteenth and early twentieth century about how China could recover its position in the world. Of the two main schools of thought that emerged in this debate, one argued that only minor changes were necessary. Western technical skills would have to be adopted, particularly in the military field, which would be used to eliminate foreign interference in Chinese affairs. China would then be able to continue on the basis of its unchanging Confucian principles. There was, however, another school of thought, whose members argued for a more revolutionary change, which would sweep the discredited old order away. What was needed, so the argument went, was a new system of belief, a new philosophy that would replace the outworn Confucianism. To some of the revolutionary school, the new Marxist-Leninist doctrine after 1917 was especially appealing. The explanation of imperialism in particular provided a convincing interpretation of the misfortunes that had befallen China and at the same time a "doctrine of certain hope."[4] Capitalism was responsible for China's problems, but capitalism was in decay and about to be replaced by the superior Communist system.

The Communist party, however, was neither the only nor the most powerful revolutionary movement in China during the first half of the twentieth century. Following the fall of the Manchu dynasty and the foundation of the republic in 1912, Sun Yat-sen's Nationalist party grew in strength until it achieved power in 1928. The clash between the two movements led to a struggle for power that characterized the Chinese political scene for much of the period between the 1920s and 1940s. Despite their major ideological differences, however, both Nationalists and Communists desired the restoration of Chinese preeminence in world politics. Both movements strove, often in uneasy alliance, to free their country from the Japanese invaders, and both envisaged the restoration of Chinese control over various territories that had traditionally belonged to China. Both Nationalist and Communist leaders have also on numerous occasions expressed similar views on China's past and future in terms of its traditional superiority and its future importance in world politics.[5]

One of the central objectives therefore of whichever power emerged triumphant from the civil war was likely to be the restoration of China to a position of some significance in the international system. As China also had not been slow traditionally to use force[6] to achieve its objectives, many predicted that a strong Nationalist or Communist government would, if necessary, use military means to recover those lands deemed to be part of China and to reestablish its position in world affairs. The victory of the Communist forces in 1949 seemed to further reinforce this view. Marxism-Leninism taught that there could be no peace between Socialist and capitalist camps. It emphasized the duty of all true Communists to transform non-Socialist countries into Socialist ones, and the only way to achieve this was through revolution. According to Mao, "Revolution and revolutionary wars are inevitable in class society . . . without them it is impossible to accomplish any leap in social development and to overthrow the reactionary ruling classes, and therefore impossible for the people to win political power."[7] China's new ideology also stressed the utility of armed force to achieve political objectives. To Mao not only did "political power grow out of the barrel of a gun," but also "with guns the whole world can be transformed."

To some observers of contemporary China in October 1949, therefore, the combination of its irredentist claims, great power pretentions, and its communist ideology was a portent of aggressive Chinese policies in international relations in the future. China could be expected to use force to recover its territories in Asia and thereby enhance its own power and extend Communist influence.

## *China's Use of Force in International Relations*

Since 1949 the debate about how far these fears have been justified has characterized much of the literature written about Chinese foreign and defense policy. Viewing exactly the same evidence different observers have come to widely conflicting conclusions. Some have argued that Chinese actions in the military field have demonstrated an inherently aggressive and expansionist policy. Others have contended quite the reverse, that China has been extremely cautious and indeed defensive in its utilization of armed force as an instrument of foreign policy.

To those who hold the former view, Chinese operations such as those in Tibet, Korea, on the Indian border, and the conflict with Vietnam as well as vocal and material support for wars of national liberation all illustrate China's aggressive intentions against her neighbors. Such actions, it is often contended, demonstrate the determination of Chinese leaders to recover all those lands regarded as being part of the old Chinese empire.

There are others who interpret the evidence available more in terms of the influence of ideology on policy-making than of traditional territorial claims. Franz Michael, for example, writing in 1971, argued that China was interested in the "far more important form of "expansionism" . . .

the spread of Communist revolution." For Professor Michael, the pursuit of international communism and the mission to spread Marxism-Leninism throughout the world have been the decisive factors in China's policies abroad. According to this view also, the acquisition of nuclear weapons by the Chinese has facilitated a much more militant policy toward Asia and a more vigorous support of wars of liberation, such that "the clear intent of current Chinese policy is to promote, foster and support Communist revolutions."[8]

Those who hold a different view of Chinese military policies since 1949 do not deny that China has constantly issued bellicose statements, made verbal assaults on the status quo and called for the overthrow of foreign governments. They accept that China has given material and moral support to revolutionary movements throughout the world. They also recognize that the Chinese leaders undoubtedly seek greater power and influence in world affairs. What they do reject, however, is that all this adds up to expansionism by the People's Republic of China. Nor are they convinced that ideology plays a crucial role in Chinese policy. Those who argue this line emphasize in particular the moderation and pragmatism of the post-Mao leadership.

In terms of assessing the validity of these two arguments, much, of course, depends on how the word *expansionism* is defined. If it is taken to mean simply an attempt to extend a nation's sphere of influence and interest, then undoubtedly China is an expansionist power. But then most states, particularly the larger states, fit into this category. Such a definition as this hardly corresponds to the usual interpretation given to the word in international relations. If, on the other hand, it is interpreted as it more commonly is, to involve "a sustained effort to exercise direct control over people and territory beyond a nation's borders,"[9] then on the evidence available it would be difficult to conclude justifiably that China is an expansionist power. As Professor Peter Van Ness pointed out in his study of Chinese military policy:

> . . . despite Peking's inflammatory rhetoric and its formidable military power, Chinese behaviour in foreign affairs has been surprisingly circumspect. In the 21 years since the establishment of the People's Republic, Chinese military units, with the exception of the Korean War, have rarely, and then only briefly, engaged in combat operations beyond or adjacent to China's borders.[10]

The validity of this view can be illustrated from an examination of Chinese military activity abroad during the period since 1949, which can be classified in three main categories: military action in support of neighboring Communist states; conflict with neighboring states; and support for wars of national liberation.

### *Military Action in Support of Neighboring Communist States*

Chinese operations in Korea between 1950 and 1953 and North Vietnam between 1965 and 1968 provide the main illustrations of the use of Chinese military forces in combat operations in territory clearly outside Chinese borders.

Chinese intervention in Korea in the autumn of 1950 and the commitment of a force of around three hundred thousand over a three-year period is often cited as a clear illustration of Chinese aggressive intention. Once again, however, the evidence does not seem to lead to the conclusion that China was behaving aggressively. As Allen S. Whiting has pointed out, China was probably not involved in the planning of the initial attack by the north on the south and only intervened when the existence of a Communist government in the north and its own security appeared to be at stake.[11] Chinese troops were only used in support of the government of North Korea after U.N. forces had crossed the thirty-eighth parallel, ignoring warnings from Peking, in an attempt to unify the country by force. From a Chinese point of view the use of force was designed to keep a neighboring Communist government in power. But perhaps more importantly, Chinese forces were committed to counter what was perceived to be a major threat to Chinese security by preserving a vital buffer zone that protected areas of critical industrial importance to the new Communist regime in Peking.

In the case of North Vietnam, Chinese troops were not in direct combat operations as they were in Korea, but their intervention does appear to have served a similar function. Chinese forces were sent to North Vietnam in 1965 presumably on the request of the Hanoi government after the beginning of the American bombing campaign against the north and the large-scale intervention of American combat forces. At no time between 1965 and 1968 were there more than fifty thousand Chinese troops in North Vietnam, and the tasks they performed were mainly engineering and construction functions, maintaining and rebuilding bridges, roads, and railway lines destroyed by American bombing. In so doing they helped relieve North Vietnamese forces for operations against the south and also to deter an American invasion of the north, which would have directly threatened the security of China itself. In March 1968 after President Johnson's decision to end the major bombing campaign, the expeditionary force returned to China. As in the case of Korea, Chinese forces withdrew as soon as they were no longer needed. It also appears that little or no attempt was made to utilize their presence on foreign soil to achieve the kind of political control that the presence of the Soviet army has made possible in Eastern Europe.

### *Conflict with Neighboring States*

China's most significant border clash occurred with India in the short war of 1962 and would seem at first glance to provide prima facie evidence for the conventional image of an aggressive China. Chinese forces initiated the first large-scale attacks of the war between 20 September and 20 November 1962 with simultaneous advances in the North

East Frontier Agency and Ladakh and swept Indian forces from all disputed territory. According to most informed commentators of the war, however, the Chinese initiative resulted from India's persistence in its "forward policy."[12] After years of passivity in the long-standing border dispute between the two countries the Indian government appears to have sown the seeds of war by sending small patrols into the unpopulated Himalayan valleys and plateaus. Although the Indian initiative posed no threat to China, it coincided with a number of other events that came to common focus for Chinese leaders in 1962. An unprecedented resurgence of Chinese Nationalist attention in "reconquering the mainland"; Nationalist and American support for Tibetan insurgents; the vulnerability of the only military road linking Sinkiang with Tibet, which traversed the Aksai Chin plateau at the Western end of the Sino-Soviet border; and problems in Sinkiang on the Sino-Soviet border all added up in Chinese eyes to the possibility of Soviet-American collusion with the Indian advances.[13] At the same time China wished to assert its claim to disputed territory and to emphasize its dominant position in Asia.

Seen in these terms the People's Liberation Army (PLA) attack on Indian forces, followed by a unilateral cease-fire and withdrawal from the North East Frontier Agency, suggests, in part at least, a limited action by Chinese leaders faced with what they perceived to be a growing threat to their security. At the same time China did hold on to disputed territory and showed willingness to use military force to achieve certain foreign policy objectives.

Another major boundary dispute in which China has used military force is that with the Soviet Union. With the deterioration in the relationship between the two states in the late 1950s and early 1960s, the disputed border territories in Sinkiang and along the Amur and Ussuri rivers flared into open conflict culminating in the major clashes of 1969. The legal position over the disputed territories and over what the Chinese describe as the "unequal treaties" imposed on China by the Russian tsars is a highly debatable one. On the evidence available, however, the Chinese appear to have been no more aggressive in pressing their demands than the Soviet Union.[14] Certainly the actions of the PLA cannot be construed as evidence of the aggressive and expansionist policies of the Peking government as the Russians would have everyone believe. Indeed from a Chinese perspective quite the reverse appears to be true. The actions of Soviet forces and the massing of between forty-five and forty-seven Soviet divisions and rocket forces along the border probably suggest to the Chinese the possibility, in certain circumstances, of a Soviet attack on China.

In the case of both the Indian and Soviet borders, therefore, it would appear that the Chinese use of force was not expansionist in the sense it has been defined. In both cases, as one commentator has pointed out,

"The location of the border line became an issue after other differences had arisen between China and the two countries, both of which had been friendly with China."[15] This seems to imply that the territory itself may not have been as important as other questions in the wider dispute that had arisen in both cases. In the same period as China was in conflict with these two states, it signed border treaties with five neighbors—Burma, Nepal, Mongolia, Pakistan, and Afghanistan—delineating common borders between itself and them. In these negotiations Chinese leaders also made generous concessions to the individual states.

Much the same can also be said about China's conflict with Vietnam in February–March 1979. After building up its forces on the Vietnam border and issuing several warnings, Chinese forces began what they described as "defensive counterattack" on 17 February. The invasion along a 450-mile front had little to do with the outstanding territorial disputes between the two countries. Rather it was designed to "punish" Vietnam for using its forces to set up and maintain a puppet regime in Cambodia (hitherto an ally of China).[16] Although the PLA achieved its major objective—the capture of Long San—the fierce resistance from Vietnamese forces made the outcome of the conflict much less decisive than similar action against India. By 15 March when China withdrew its forces both sides had lost about three thousand killed and more than twenty thousand wounded or taken prisoner. The Chinese believed that they had made their point that Vietnam could not act with impunity in Southeast Asia and had shown that it was prepared to use force in support of an ally even if by so doing it risked conflict with the Soviet Union. As such China probably felt that it had reasserted its claim that Southeast Asia was very much part of what it regarded as its sphere of influence.

At the same time the costs in material and political terms were clearly very great. Apart from the high casualties in such a limited incursion, the deficiencies shown up in the PLA and its inability to secure an outright victory could not have been lost on the Vietnamese or the Soviet Union. Vietnam continued its policies in Cambodia (Kampuchea), and the Vietnamese-Soviet alliance was further strengthened, thus challenging Chinese claims to supremacy in the region. In many respects therefore the Chinese were much less successful in achieving their objectives than they were in the Sino-Indian War of 1962.

In each of the three cases discussed the PLA was not used specifically to resolve border disputes and acquire territory (although these were considerations in the Sino-Indian war). Military force tended to be seen more in political terms to suggest to a neighbor that China could not be pushed around. The willingness to fight and to take the conflict into the enemy's territory was designed as a forceful signal to the lengths China would go to defend its interests in the wider political disputes with the countries concerned. The use of such force also seems to have been an attempt to deflate the pretentions of each of the three countries involved to assume predominance in regions around the periphery of China. In so doing China was asserting her own choice to regional influence.

An appraisal therefore of all of the cases of direct Chinese military activity in border disputes and assistance to neighboring governments reveals a pattern of military involvement for limited objectives. In all of the operations cited, after hostilities had ended Chinese forces were largely withdrawn. Apart from Chinese occupation of Aksai Chin there was "virtually no gain for China in terms of expanded territory or political control as a result of its military actions."[17] It must be said, however, that although Chinese actions were not unduly aggressive, neither were they wholly defensive. In each of the cases the Chinese showed no reluctance to use military force in a pragmatic and calculated way as an instrument of foreign policy.

*Support for Wars of National Liberation*

Those who support the thesis that China has been rather more aggressive than defensive argue that "Peking's disinterest in old-style empire-building has deflected Western attention"[18] from the indirect form of Chinese expansion, which has taken the form of the support for foreign revolutions. According to this argument, Chinese leaders, in the past, have been engaged in fostering revolution by providing moral and often material support for individuals and organizations engaged in making revolution against their governments. The validity of this view would seem to be strengthened by the wealth of evidence available on Chinese assistance to insurgent movements all over the world. It is often pointed out that Chinese leaders have supported many revolutionary groups by offering them training facilities and instruction in China, weapons, and financial assistance.

The Chinese have made it plain, however, and past experience would seem to reinforce their public statements, that revolutionary movements must rely essentially on their own efforts and resources to gain power. This concept of "self-reliance" is, from a Chinese viewpoint, one of the most important tenets of Maoist revolutionary doctrine. In his famous, but often misunderstood, article entitled "Long Live the Victory of People's War" written in 1965, Lin Piao spells out the Chinese view that revolution cannot be exported and that insurgent forces must be almost entirely self-sufficient if they are to succeed.* This contention would seem to be confirmed by Lovelace's empirical study of Chinese policies towards insurgent movements in Thailand. According to this study the Chinese, in this particular case, lived up in practice to the principle of self-reliance. Lovelace concludes that Chinese support was "certainly not a recipe for direct CPR [Chinese People's Republic] involvement or control of revolutionary movements."[19]

---

*Some have wrongly interpreted this article as a blueprint for worldwide revolution.

Once again, therefore, it would seem that the picture of an aggressive China supporting and controlling revolutionary movements throughout the world as part of Peking's grand design does not conform strictly to reality. China does attempt to influence the internal politics of other states from time to time by providing support for those domestic forces whose policies it favors. In this, however, it is perhaps like virtually all the other major powers. As with the other examples of the more direct use of force by the People's Republic, the limitation on Chinese support of wars of national liberation, stemming from the concept of self-reliance, would seem to further confirm the contention that while China has been perfectly willing to use force in defense of its interests, it would be difficult on the basis of the evidence available to describe her as an expansionist power in the period since 1949.

## *The Use of Chinese Forces in the Domestic Context*

Besides the limited use of force as an instrument of foreign policy, the Chinese leaders have also used their armed forces for a variety of specific domestic purposes. In fact a strong case can be made that the PLA is seen by the leaders in Peking as being just as important in the domestic as in the international context. At times it has probably appeared even more important. The domestic purposes that the armed forces have served range right across the spectrum from the direct and physical use of force to reincorporate traditional Chinese territories into the new Communist state to the peaceful use of the PLA in a whole host of roles in support of the Chinese economy and in domestic political disputes.

### *The Physical Use of Force in the Domestic Environment*

Following the Communist victory over the Nationalist forces in 1949, the PLA had a number of limited, but nevertheless pressing functions to perform internally. These, according to Liu Shao-chi, speaking in May 1950, included the occupation of various territories considered to be an integral part of China.

Among the first tasks of Chinese forces was the occupation of Tibet and the invasion of Taiwan. On 2 September 1949 the New China News Agency (NCNA), circulated an article entitled "Foreign Aggressors Must Not be Allowed to Annex Chinese Territory," in which the determination of the new regime to "liberate all Chinese territory including Tibet, Sinkiang, Hainan and Formosa" was made absolutely clear.[20] By April 1950 Sinkiang and Hainan had been seized, leaving Tibet and Formosa on the list of unfinished business. Preparations for the occupation of Tibet began in January 1950 as soon as southwest China had been liberated. In October a force of forty thousand men crossed into Tibet and captured the town of Chamdo in the east in what was to be the first and last major engagement of the invasion. By 23 May 1951 an agreement had been reached with the Tibetan authorities for "the peaceful liberation of Tibet," and by February 1952 the Military Region of Tibet had been established. For the Communist Chinese, Tibet, which had never been generally recognized as an independent state, was within what was considered, even by the nationalists, as the proper limit of the Chinese domain. The Communists felt therefore not only justified in taking such action but also confident that outside powers would not intervene on behalf of the Dalai Lama.

The new Chinese leaders held the same expectations about the invasion of Taiwan. The determination to undertake the invasion, which was to be "the principal task" of 1950, was further reinforced by President Truman's announcement on 5 January 1950 that the United States recognized China's legal claim to Formosa. He also promised that the American government had "no intention of utilizing its armed forces to interfere in the present situation."[21] Preparations for the invasion, however, by General Ch'en Yi's force of about three hundred thousand were frustrated by the outbreak of the Korean War and the radical change in U.S. attitudes. Following the invasion by North Korea, President Truman announced on 27 June that the Seventh Fleet had been ordered to the Taiwan straits to prevent any attack on Formosa. It was at this point, according to Whiting, that Peking tacitly admitted the "futility of attacking Taiwan as long as the threat of American intervention remained."[22] Since then no really determined efforts have been made to take Taiwan through the use of armed force. Even the two Taiwan Straits crises of 1954–55 and 1958 must be viewed more in terms of China's wider political objectives* than as a serious attempt to overthrow the Nationalist government on the island.[23]

The Chinese claim to Taiwan, however, has remained an important aspect of policy ever since the 1950s. For many years it was a serious impediment to efforts to improve Sino-American relations. For China it remained a piece of lost territory that had to be reincorporated into China in the long run either peacefully or through the use of force. For the United States, on the other hand, Taiwan was an ally and there could be no reunification with the mainland except through negotiation. During the discussions on normalizing relations between Peking and Washington in 1978 both sides agreed to shelve the issue for the sake of their wider interests. Since then it has remained a central issue for the Peking leaders although diplomatic offensives in recent years have emphasized a peaceful solution to the problem. In line with this China has made a number of attempts to open discussions with Taiwan on the peaceful reunification of the island with mainland China. In its nine-point proposals of September–October 1981 Peking offered Taiwan the right to have its own armed forces and economic system, the opportunity to

*China, in part at least, initiated the crises, particularly in 1954, to test the alliance with the Soviet Union.

maintain its trade and cultural ties with other countries, and even high positions in a reunified China for senior Taiwanese officials. The offer was, however, rejected by Taiwan.[24] In August 1982 also the Peking—Washington agreement on arm sales to Taiwan was followed by offers of reconciliation from China to Taiwan. Once again it was turned down by the Taipei leadership. Significantly, despite bending over backwards for a peaceful settlement of the issue, throughout this period when diplomacy has been to the fore, the Chinese leaders nevertheless refused to give up the option to use force if that should ever prove necessary.

### *The Peaceful Uses of the PLA*

Apart from the more traditional use of armed forces in support of both foreign and domestic policy, the PLA also holds a unique place among the world's armies in terms of the role it has played in the economy and more generally in the political system.

The role of the PLA during the civil war in supporting the peasantry in various forms of production was carried on after the Communist victory, becoming at times even more important than the more military functions of the army. The army's work in this field ranged from agricultural production, water conservation, and flood control to industrial production and management. Such activity reached a peak during the Great Leap Forward period when, according to Chinese sources, the army contributed respectively 59, 44, and 46 million man-days of labor to the economy in the years 1958–60.[25]

As well as this important economic function, the PLA has also at various times played a critical political role. In particular, the army and militia have worked, especially in the late 1950s and 1960s, to politicize Chinese society. The "Everyone a Soldier Movement" launched in the autumn of 1958 during the first months of the Great Leap Forward and the "Learn from the PLA" movement in 1964 were both attempts to use the army and militia as vehicles of political education. In the case of the latter campaign, the PLA was held up as a revolutionary model for Chinese society. According to the slogans of the campaign the army was worthy of emulation because its working style had enabled it to achieve the Maoist goal of being both ideologically correct and professionally competent.[26]

Apart from its role in political socialization the PLA has also found itself involved from time to time in maintaining domestic order. The most notable example of such action came in the late 1960s during the Cultural Revolution. In some areas of China the fervor and excesses of the Red Guards brought civil strife and chaos, which the army was called upon to suppress. In so doing the PLA, with its professional interest in law and order, found itself in conflict not only with the Red Guards but also the more radical members of the Chinese leadership, who saw the Cultural Revolution as an opportunity to further their own personal and ideological positions. Traditional suspicion between senior military figures and "radical" leaders was transformed into deep-seated hostility during the late 1960s and in the final years of Mao's life. As such in the unsettled political period between 1971 and 1976 the military leadership found itself increasingly prominent politically in alliance with "moderate" political leaders against the Gang of Four and their supporters. With Mao's death in 1976 and the absence of the chairman "to hold the snarling rivals in check," the PLA played a crucial role in supporting the moderate faction under the Hua Kuo-feng to defeat the Gang of Four.[27] Such decisive intervention not surprisingly left the PLA in a strong political position in China. Although its leaders were concerned to return to barracks after the succession struggle, the army has nevertheless retained a strong voice in the central decision-making organs in the 1980s.

### *"Red" versus "Expert"*

The struggle with the "radical" leaders in the late 1960s and early 1970s was reflected in the major debate between "revolutionization" and "professionalization," which had its origins in the primary function of the PLA after victory in 1949. In the civil war, the PLA was the nucleus of the revolution. Its tasks were difficult and demanding but in many ways were relatively uncomplicated. The army played its part in the political struggle of winning popular support and performed its military function of wearing down and defeating an identifiable opponent. Once victory had been achieved, however, the functions of the PLA changed dramatically. Now, instead of being a force designed solely to overthrow the established authorities, it became itself an instrument of government policy with a wide spectrum of international as well as domestic objectives to fulfill. As John Gittings aptly puts it, "By the very act of victory, the PLA had transformed itself from a free-wheeling revolutionary army to a predominantly garrison army of national defense."[28]

Although the initial steps of military reform were laid down in the Common Program of the People's Political Consultative Conference in September 1949,[29] the real impetus for modernization first became apparent during the Korean War. Despite the initial Chinese successes, later setbacks soon demonstrated the inherent weakness of the army engaged in action against a sophisticated opponent using the advantages of modern technology. Without the benefit of indigenous armaments industries, the obvious source of modernized equipment and techniques was China's new ally, the Soviet Union, with whom she had signed a treaty of Friendship, Alliance and Mutual Assistance in 1950. The gradual reliance on the Soviet Union for weapons systems led also to the attempt to emulate the Soviet model in terms of military organization as well as strategic doctrine. A system of formal ranks was introduced, and specialized service branches were created in the PLA similar to those in the Soviet army. Conscription was also adopted on Soviet advice as a means of securing the necessary quality of recruits who would be capable of learning the technical skills associated with a modern military force. According to the editorial of the *People's Daily* in July 1954, "The Soviet Army provided the PLA with a great example, the future PLA would resemble the Soviet Army of Today."[30]

The transformation of the PLA from a revolutionary guerrilla force of volunteers to a mixed conscript-professional army soon brought some sections of the new officer corps into conflict with the party leadership. Growing impatience with political control and pressure for an accelerated program to improve the Chinese military posture led to conflict in the mid-1950s. The military modernizers, however, were faced with a hostile coalition of other military and civilian interests that desired an acceleration of economic development through the reduction of the military budget. The antimodernization lobby seems to have been strengthened at this time also by a growing disenchantment, which was emerging in some sections of the party, with the Soviet model in both economic and military spheres. Increasing questioning of Soviet practices and growing economic difficulties requiring concentration on the problems of reconstruction led in April 1956 to a partial reaction against the program of military modernization in the "Ten Great Relationships." Mao agreed that modern armed forces were necessary but declared that military expenditure must be reduced rather than increased to facilitate an acceleration of economic development.[31] The pursuit of a modernized army remained an important objective, but equal priority was given to the army's nonmilitary functions in the spheres of production work and politicization, and political control over the army was stressed.

The consensus that the 1956 compromises attempted to achieve between professionalization and revolutionization gradually broke down however, in the following two years. Many professional combat officers became more and more critical of government policy as less stress was placed on modernization and more on production work, politicization, and political control. The increasing emphasis on domestic functions also came at a time when serious strains were appearing in the Sino-Soviet alliance. Soviet support for Chinese actions in the Quemoy crisis of 1958 came only after China had borne the greatest risks. Further questioning of the Soviet military model that this caused seems, in large part, to have been an important contributary factor in the readoption by the Chinese Communist Party (CCP) of the traditional Chinese revolutionary model, which had been so successful before 1949. The concept of people's war was given renewed emphasis, and the militia was strengthened in an attempt to improve China's defense posture without reducing the army's participation in domestic political and economic projects and without reducing the funds set aside for the civilian economy.

From the professional military point of view, however, the militia was no substitute for highly trained and well-equipped armed forces. To many senior officers the renewed stress on people's war and the concern with nonmilitary roles was seriously detrimental to the fighting capability of the Chinese army. The articulation of their objections to the changing emphasis in military policy dictated by the party led, however, to the purging of the minister of defense, P'eng Teh-huai, and some of his colleagues in 1959.[32] Once again the supporters of modernization and professionalization seemed to have been defeated by those who wished to preserve the revolutionary character of the army. Although this was true to a certain extent, as in 1956, the solution to the crisis in civil-military relations was more of a compromise than it seemed. While stressing the continuing priority given to the domestic role of the PLA, P'eng's successor as minister of defense, Lin Piao, also continued the task of modernizing the armed forces. He intensified political training in the PLA and rejuvenated party branches throughout the military hierarchy, but he also devoted considerable effort to the development of nuclear weapons as well as the creation of new machine-building ministries charged with the production of sophisticated military weapons.

A similar set of grievances followed by a compromise solution that nevertheless largely reasserted revolutionary principles occurred in 1964–65. The decision to use the PLA to a much greater extent as the focal point of the Great Cultural Revolution and the intensification of the Vietnam War led military leaders like Lo Jui-ching, the chief of staff, once again to emphasize the professional qualities of the army. Lo Jui-ching urged that less emphasis should be given to the PLA's domestic roles and improvements should be made in China's military preparedness.[33] To the Chinese chief of staff, the threat of an American attack on China following its direct involvement in Vietnam meant that there were "a thousand and one things to be done."[34] Among the most important of these was the task of improving the size and effectiveness of the conventional standing forces. In the new circumstances, what was needed, according to Lo, was a new strategic posture in which the emphasis should be placed less on the people's war concept and more on the professional option of a nationally coordinated, technologically sophisticated defense.

The public criticisms of Lo's suggestions concentrated on the question of enemy intentions and the efficacy of the reliance on the people's war concept. In opposition to Lo, Lin Piao questioned his estimate of the likelihood of an American attack on China and called for an increased rather than a decreased reliance on people's war. In private Lo's campaign for greater modernization became inextricably linked with the struggle for political power, such that his proposals were deliberately exaggerated by the coalition of bureaucratic interests that opposed him. According to W. W. Whitson, "Lo was accused of concocting an elaborate rationalization for proposals whose real purpose was to force the disengagement of the PLA from political activity, reduce its political reliability and thus encourage the spread of revisionism in China."[35]

Although Lo Jui-ching, like P'eng Teh-huai, was purged, the solution that followed his dismissal, as on previous occasions, was again something of a compromise. Predictably increased emphasis was placed on the political roles of the PLA. But, while many of Lo's proposals were rejected, some were nevertheless accepted. His suggestions, for example, that defenses in the south should be reinforced and that aircraft production should be expanded were both put into effect. Despite the stress on the revolutionary characteristics of the PLA also during the Cultural Revolution and the increasing role of the army in political affairs, the modernization of the navy and the priority given to nuclear testing were also continued with little interference.

A similar pattern of gradually continuing the process of modernization of the PLA with the parallel stress on revolutionary principles has characterized Chinese defense policy since the fall of Lin Piao in September 1971. Although the issues at stake in the purging of Lin Piao were far more complex than simply professionalism or revolutionization in the PLA, among the chief concerns of the anti-Lin coalition, led by Chou En-lai, was the determination to reimpose party control over the army. In some ways Chou En-lai's "New Course," which emerged after Lin Piao's fall from power, represented an attempt to get away from the more extreme revolutionary manifestations of the Cultural Revolution. It did, nevertheless, signal an attempt to ensure that "the party controls the gun." Following Lin's demise, a concerted nationwide campaign was undertaken to downgrade the role of the military and to reassert party control. Evidence exists that altogether fifty-six leading military figures disappeared in the purges that followed Lin's fall, five of whom were Politburo members.[36]

Differences also apparently existed, as on previous occasions, over attitudes toward the Soviet Union. Some evidence exists that Lin Piao and some sections of the professional military machine were opposed to the designation of the Soviet Union as the primary threat and the proposed limited rapprochement with the United States.[37] From the military viewpoint closer American relations meant the possibility of antagonizing Moscow and therefore increased the danger of an unwelcome Sino-Soviet war. It also further reduced the opportunity for patching up the differences with the Soviet leaders, which would have enabled the PLA to return to the kind of military cooperation characteristic of the 1950s.

The period between Lin Piao's demise in 1971 and Mao's death in 1976 was a period of indecision in China, as the power pendulum swung back and forth between the "radical" and more "moderate" factions of the leadership. In its attempts to reassert its professional interests after 1971 the military suffered from the suspicion created by Lin's attempts to use the PLA to reinforce his own political ambitions.[38] The military also found itself under constant pressure from the radicals to upgrade political activities within the armed forces. As such, in the twilight period of the Maoist era the climate of acute antiprofessionalism that prevailed brought a significant deterioration in the combat effectiveness of China's armed forces. Red rather than expert tended to hold sway.

The defeat of the Gang of Four, however, and the ascendancy of the military backed by a moderate civilian leadership after Mao's death not surprisingly ushered in a new era of professionalism in the PLA. With less need to involve themselves in political administration, the armed forces have focused more on modernization and professional considerations since 1976. This preoccupation has been encouraged by the civilian leadership, which has recognized the need to raise professional standards and modernize the weapons and equipment of the PLA. As a result, renewed emphasis was given in the late 1970s to air defense, armaments production, and research into military technology; military delegations have been sent abroad to look at foreign military technology; and military delegations from Western states to China have been sent to improve both military discipline and military training in the PLA.

Alongside this return to professionalism has gone a resubordination of the army to the party. Political control has, however, been exerted in a more moderate and sensitive way than in the past so as not to infringe on the professional requirements of the armed forces. This has meant, however, that the army has not been able to have it all its own way in the battle for resources in China since Mao's death. The debate from 1976 to 1978 between those who wanted rapid modernization of the PLA's weapons and those advocating a more gradual modernization was resolved in early 1978 largely in favor of the latter.[39] Military modernization would take place, but as Hua Kuo-feng indicated in the Report to the Fifth National People's Congress in February 1978, it will follow, not precede, wider economic development. As such, the expert as opposed to the red viewpoint is now in the ascendancy in China, although as Joffe and Segal have put it, "As long as the current program of national development remains in effect, then the Chinese army will move toward comprehensive technological modernization, but its march in that direction will apparently be long and slow."[40]

## *Alliances and National Self-Sufficiency*

The domestic debate both within the PLA and the party and between sections of both over the question of professionalism and modernization was linked very loosely, as has been suggested, with differing ideas on the value of alliances, particularly that with the Soviet Union.

In 1949 the creation of a new Communist state in a hostile world at the height of the cold war seemed to suggest to the leadership in Peking that an alliance with the Soviet Union made sound political, military, economic, and indeed ideological sense. For some time following the signing of the treaty with Russia in 1950, China's strategic policy was based on the assumption that the United States was the major threat and that the Soviet Union would remain a reliable ally. Reliance on the Soviet Union was particularly important in terms of the supply of the much-needed military hardware and the nuclear guarantee that Russia was also able to offer. The benefits of the alliance became particularly clear in October 1957 with the agreement signed by the two governments, called "New Technology for National Defense." According to subsequent Chinese sources, the agreement included a Soviet promise to "provide China with a sample of an atomic bomb and technical data concerning its manufacture."[41] The Russian leaders also supplied China with an experimental reactor and undertook to train Chinese atomic scientists at the Joint Institute for Nuclear Research at Dubno in the Soviet Union. With the help of Soviet military aid, therefore, China began to develop not only its own conventional defense forces, but also took the first tentative step toward the production of a nuclear force.

Despite the military advantages to China of the Soviet alliance, the seeds of future discord appear to have existed from the beginning. Even during the two and a half months of negotiations that resulted in the Sino-Soviet Treaty of 1950, Mao did not obtain the kind of satisfaction he expected. To be sure, according to the treaty Soviet rights in Manchuria were to be surrendered, but not for three years; and Soviet credits to the new state of $300 million were hardly overgenerous given the size of China and all the difficulties of reconstruction it faced. Soviet interests were also to be perpetuated in Sinkiang by means of the far-from-equal joint stock companies.

Difficulties in the alliance also seem to have occurred in the first year of the Korean War. Although China's intervention in the war was eventually to open the floodgates of Soviet aid it would appear that the assistance was not at first forthcoming on anything like the scale the Chinese expected. In his analysis of the Sino-Soviet relationship at this time, John Gittings detects "a degree of coolness towards the Soviet Union which only diminished as Soviet aid became available on a more generous scale." Despite more substantial aid later in the war, the Korean experience seems to have had a significant impact on the Chinese attitudes toward her ally, particularly in retrospect. In later Chinese polemics with the Soviet Union, the leaders in Peking have clearly demonstrated their dissatisfaction with the way they were treated by the Kremlin during the war. In a letter from the Central Committee of the CCP to the Central Committee of the Communist Party of the Soviet

Union (CPSU) in May 1964, the Chinese claimed to have shouldered heavy sacrifices and stood in the first line of defense in Korea "so that the Soviet Union might stay in the second line."[42]

Chinese uneasiness over too great a reliance on the Soviet Union, which was a product of the gradually deteriorating relations during the 1950s, came to a head following the second Taiwan Straits crisis in 1958–59. It would appear from the limited evidence available that the Soviet Union attempted to get the leaders in Peking to accept proposals for increased integration of the Chinese defense effort with other Communist bloc countries. At the same time as Marshal P'eng Teh-huai negotiated with Soviet officials on questions such as these, however, the general trend in Chinese foreign policy was in the opposite direction, toward greater "self-reliance" symbolized by the Great Leap Forward movement.[43] Despite the opposition of those like P'eng Teh-huai, greater emphasis was increasingly being placed on Chinese solutions to Chinese problems and the inadequacy of reliance on other countries in the economic, political, and military fields.

The questioning of the value of the alliance with the Soviet Union and the growing stress on national control in defense was felt particularly in nuclear technology. During the 1958 negotiations, the Russians appear to have backtracked slightly in their 1957 commitment and refused to supply the Chinese with nuclear weapons or technology unless they retained effective control over these weapons. To many Chinese leaders such Soviet control over China's nuclear weapons, together with the proposals for increased defense integration, would have made the Chinese almost totally dependent on the Soviet Union for their security. There was growing evidence of the unreliability of the Soviet leaders to pursue policies in line with Chinese interests, and to a nation as proud as China this was something that the leaders in Peking were not prepared to accept. Soviet attempts at accommodation with the West at a time when "the east wind prevailed over the west wind" convinced many Chinese leaders that the Russian guarantee was of little value. Even before the Soviet government therefore had unilaterally torn up the 1957 agreement in June 1959, the Peking leadership had already decided to develop an independent deterrent of its own.[44] The pressing need for such a force under its own control was clearly spelled out by Chen Yi in an interview with an Australian film director, John Dixon, in September 1963. In answer to the question of why China needed to construct its own nuclear force when the Soviet Union had given assurances about the defense of China against foreign aggressors, Chen Yi replied: "In the first place, what is this Soviet assurance worth? . . . How can one nation say that they will defend another—these sorts of promises are easy to make, but they are worth nothing. Soviet protection is worth nothing to us."[45]

The Chinese view of the utility of nuclear weapons is thus in many ways similar to that of the French. For the Chinese in their search for equality with the United States and the Soviet Union, nuclear weapons are the supreme symbol of national self-reliance. In a comparative study of the Chinese and French nuclear programs, B. W. Augenstein has shown that statements by the leaders of the two states reveal a high degree of similarity in their positions. Both see the development of national nuclear forces as the "marks of national greatness, and political power and importance."[46] In both cases also the possession of nuclear weapons has been seen as an important symbol of sovereignty and of a certain reluctance by either government to rely on alliances for its national security.

For much of the 1960s China held a "three worlds" concept, which envisaged the Soviet Union and the United States on one level as aggressive superpowers that would inevitably go to war one day; the other industrialized nations as on another level; and the undeveloped or developing world (including China) as on the third level. According to this concept it was necessary for the Third World countries to join together to overcome the threat posed by both superpowers. During the 1970s this thesis was largely abandoned as China and the United States moved toward a normalization of relations. In a discussion of Mao's "Theory of the differentiation of the Three Worlds," published in the *People's Daily* in November 1977, the Soviet Union was firmly identified as the major source of war. The United States was viewed, on the other hand, as only trying to retain its position in the world.[47]

Another significant modification in Chinese thinking came subsequently with the abandonment of the idea that world war was inevitable. This was replaced in the late seventies by the notion that war could be averted by the formation of a "united front." Significantly during his 1979 visit to the United States Deng Xiaoping talked about a "united front," including the United States, which would oppose the hegemonist policies of the Soviet Union. Deng also talked about a "new strategic partnership" with the United States.

The concept of a "united front," however, should not be confused with an alliance. In the words of William R. Heaton, Jr., the

> "united front" is a shifting coalition that involved all forces that can be involved at a given time. As the nature of the threat changes, so must the nature of the "united front." The "united front" is not an alliance, but an expedient policy based on self-interest.[48]

In this respect the Chinese see the concept of a "united front" as being much looser and less permanent than traditional alliance relationships.

Chinese skepticism over alliances (dating from their experiences in the 1950s and 1960s) became very clear in the early 1980s when renewed difficulties over the Taiwan issue brought a cooling in Sino-American relations after the growing collaboration of the 1970s. Although the question of Taiwan had not been resolved, it had been shelved by the joint normalization communiqué of 1978. However, the wavering of the new Reagan administration over its China policy and the proposed sale of the F-X fighter to Taiwan brought a significant deterioration in Sino-American relations in 1981. Such was the dismay in Peking that toward the end of the year the United States was being characterized in the Chinese press as "a superpower competing with the Soviet Union for hegemony over the world"—a line that had been promoted while Mao was in power but that had been dropped from 1979 on as China moved closer to the United States.[49]

Secretary of State Haig's statement that the 1978 normalization treaty remained the basis of America's China policy and the decision in January 1982 not to sell the advanced fighters to Taiwan helped to a certain extent to improve the atmosphere of Sino-American relations. The Sovet Union was also somewhat slow to exploit the difficulties although in early 1982 Brezhnev did call for an improvement in relations between the Soviet Union and China and a resumption of the border talks (interrupted by the Soviet invasion of Afghanistan). A tentative attempt to overcome the outstanding difficulties also characterized the early part of the Gorbachev period. At the time of writing there would seem to be some kind of strategic reassessment being undertaken in Peking. Relations with Washington clearly remain closer than with the Soviet Union. The lesson of the Taiwan issue, however, should not be lost on the United States, particularly in terms of the self-interest and expediency with which China clearly views its policy of a "united front."

## *China's National Defense Doctrine: The People's War Concept*

The Chinese view of alliances has had a significant impact upon the defense strategy they have evolved. The perception of a threat from the United States together with the refusal to compromise with the Soviet Union in the late 1950s and early 1960s meant that Chinese leaders deliberately accepted that they would render themselves temporarily more vulnerable to the danger of enemy attack. Without Soviet assistance and without any form of nuclear deterrence in the short term, the Chinese leaders had the task of developing an effective national security doctrine initially against the perceived American threat and subsequently against what became the dual threat from both the "imperialists" and the new "social imperialists" in the form of the Soviet Union. Given the perception of the threat, and the historical and ideological perspectives of Chinese leaders, China's vulnerability and weakness dictated a defensive and deterrent strategy based on the people's war concept.

Mao recognized that if the "imperialists" attacked, they might use nuclear weapons as well as bacteriological and chemical agents. In order to defeat China, however, such an attack, according to Mao, would have to be followed up by an invasion of ground forces. It was here that the supremacy of the Chinese people's war concept would be felt. Such attacks, Mao maintained, would be countered by "a protracted broken-backed" war supported by the "aroused masses" of people and fought by the large regular forces, local forces, and massive militia developed under the concept of "everyone a solder." Protracted war would begin with defensive operations, to be followed by a period of stalemate and finally a conventional and positional offensive phase. In the third of Mao's phases mobile conventional operations would be combined with widespread independent guerrilla warfare.[50]

The idea behind Maoist strategy was to trade "space for time" and to achieve "active defense in depth." The task of the PLA forces was to lure the enemy deep into Chinese territory. In this way, hostile forces would be "bogged down in endless battles and drowned in a hostile human seam."[51] The defeat of the invading forces would thus be achieved by "wearing them down, defeating them piecemeal and finally driving them from the country,"[52] in the same kind of operations that had proved so successful against Chiang Kai-shek's forces. It was perhaps the prospect of fighting such a protracted indecisive campaign that discouraged the Soviet leaders from attacking China in the periods of more extreme hostility since 1969.

As well as being a defensive strategy, therefore, Mao's concept of people's war is also, and perhaps more importantly, part of China's deterrent strategy. In the period of conventional and nuclear weakness, the Chinese leaders appear to hope that the declaration of their determination to fight a long, drawn-out guerrilla campaign will deter potential enemies by convincing them that the attempt to conquer China would be too costly.

Perhaps it would, however be wrong to view such a strategy simply as a "doctrine of necessity" as some have done.[53] The inadequacies of the Chinese military machine notwithstanding, the vastness of China and the size, experience, and morale of the PLA could well prove as decisive as Mao claims in a conventional or even a limited nuclear battle on Chinese territory. Certainly before initiating an attack on China, an opponent, even one with a greatly superior military establishment like the Soviet Union, would have to think very seriously about the possibility of having large numbers of troops tied down in a long, indecisive campaign. An opponent would also have to consider the effect of such a campaign on its military forces in relation to its other major adversaries. In this sense, the deterrent value of the threat to use people's war tactics might well be high.

Since the purge of the Gang of Four there has been a tendency for Chinese officials to refer to "people's war under modern conditions."

This has been taken by some Western analysts to indicate that China is moving away from the classic people's war concept. Writers like Jack Harris and William R. Heaton, Jr., have suggested that the modernization of the PLA and the stress in Chinese defense policy debates on positional warfare in recent years heralds a significant shift in Peking's military thinking.[54] Those who support this view often quote a speech made by Defense Minister Xu Xiangqian in August 1978 on the fifty-first anniversary of the founding of the PLA. Outlining Chinese military strategy Xu argued that:

> Chinese forces will defend key points, will prevent enemy forces from moving into Chinese territory unchecked, and will systematically lead them to battlefields of Chinese choice and wipe them out piecemeal. . . .
>
> China will continue to modernize its armed forces and improve training and equipment. It will pay attention to the experience of other countries and utilize all resources for improving the capabilities of the armed forces.[55]

The stress on defending fixed points and the importance of modern weapons is seen as being alien to the traditional Maoist ideas of flexible guerrilla warfare and the emphasis on the human element in war.

It would seem, however, that such an interpretation misrepresents both traditional Maoist ideas and "people's war under modern conditions." Mao's military strategy—despite the stress on guerrilla warfare in the early stages of a protracted war—envisaged (as we have seen) a final conventional and positional phase. The mobile character of Chinese armed forces was to be maintained in the third phase, but Mao also realized the importance of positional warfare. As Gerald Segal has written, in the years after Mao wrote his ideas down, China became an established state with borders to defend and therefore had somewhat different priorities than did the revolutionary leadership engaged in a first-phase strategy.[56]

As a result much greater attention was paid in the 1960s and 1970s to the final phase of positional defense. This was not a rejection of earlier strategic ideas but an attempt to adapt them flexibly to the circumstances prevailing. In essence, in a more specific way, this is what Chinese strategists have done in recent years.

"People's war under modern conditions" does involve important improvements in Chinese forces (as we have seen). The overall principles of Chinese strategy, however, would seem to remain much the same. The enemy is to be lured deep into China. In the early stages key cities and military bases are to be defended while guerrilla forces harass the invaders' lines of communications and rear echelons. Gradually as the enemy weakens through constant attrition the PLA will then move from positional defense to positional offense to defeat the enemy in conventional pitched battles.

In their scenarios for countering invasion the Chinese recognize that they will be "pitted against superior military technology and better arms."[57] Modernization of the armed forces is therefore required, but the stress remains on the importance of what are seen as traditional Chinese natural advantages in terms of geography and popular involvement. As Xu Xiangqian pointed out in the 1978 speech quoted earlier:

> China must continue to make the most of its vast territory and large population, the superiority of its social system, the glorious tradition of people's war, and other favourable conditions to win victories through arduous and protracted people's war. . . .
>
> If any enemy dares to invade China the main forces of the country and local corps will annihilate large numbers of enemy troops. The militia will be mobilized to look for any openings to hit the enemy. Everyone will be a soldier, every village a fortress and every place a battlefield. . . .[58]

Although Mao's military principles have been adapted in a reasonably flexible way by contemporary Chinese strategists, "people's war under modern conditions" does not appear to be an abandonment of those principles.

## *Chinese Attitudes toward Nuclear Weapons and Deterrence*

It was often argued in the past that one weakness in Chinese strategic thinking was a misunderstanding of the significance of nuclear technology. Those who took this view pointed to Mao's celebrated image of the atomic bomb as a "paper tiger" and his emphasis on the superiority of "men over weapons." Mao is said to have declared that "weapons are an important factor in war, but not the decisive factor; it is people, not things, that are decisive." Some commentators have argued that this tendency to underestimate nuclear power has resulted in a rather irresponsible and cavalier attitude on the part of Chinese leaders to nuclear war. This theme, perhaps not surprisingly, is one that the Russians have often developed. Soviet sources quote Mao as saying that even if half the population of the earth were destroyed in a nuclear war, this would not be so bad because such a war would destroy capitalism and leave socialism victorious. Chinese attitudes toward disarmament and arms control, particularly the refusal to sign the Test Ban Treaty and the Nonproliferation Treaty, are also frequently cited as evidence of Chinese irresponsibility.

This interpretation of the Chinese view of nuclear weapons is widely misleading and largely incorrect. Careful analysis of what the Chinese say and what they have done in various crisis situations reveals that they have been very keenly aware of the disastrous implications of a thermonuclear war for China and for the world.[59] To say that atomic weapons are "paper tigers" does not mean, from the Chinese viewpoint, that they are powerless. It means that the Chinese believe that they are not as powerful as they appear to be. Contrary to the Soviet view in the late 1950s, the Chinese believed that it was possible to support wars of national liberation without such wars escalating into nuclear conflict. In this they would seem to have had a more realistic appreciation of the political limitations of nuclear weapons than the Russians.

One of the reasons for the misunderstanding of Chinese attitudes toward nuclear weapons is that China itself has been reluctant to develop its ideas publicly on the subject. However, the vagueness of Chinese theories on nuclear deterrence obviously does not necessarily mean that the Chinese were unaware of the realities of nuclear power. On the contrary, it would seem that China for a long time was pursuing a policy of "calculated ambiguity"[60] in its strategic doctrines to maximize opponents' uncertainty about Chinese intentions and capabilities. The value of such an approach is particularly evident for a power like China engaged in the laborious process of trying to develop an effective deterrent against vastly superior opponents. For a long time China was without an effective second-strike force, and as such the credibility of its deterrent remained in grave doubt. The belief by other powers that China's leaders were irresponsible and the tacit encouragement of that belief by China served to obfuscate this weakness. The development of quick-reaction techniques and early-warning capabilities by the Chinese helped in a way to enhance this uncertainty about China's likely reaction in a crisis. Such capabilities appeared to China's opponents, rightly or wrongly, as the basis of fire-on-warning tactics. As Harry Gelber pointed out in 1973, "such tactics may involve grave dangers and instabilities, but China's opponents cannot assume, especially in the absence of clearly declaratory policies or of a strategic dialogue between Peking and the outside world, that China might not adopt them." A nuclear strike by China's limited nuclear force against the cities of a superior enemy or that enemy's allies might not be a credible policy for a rational power, but there can perhaps be no assurance, particularly for the Russians, that an "irresponsible" Chinese government might not adopt such a policy under certain circumstances. In this sense it is possible to speculate that the Chinese reluctance to develop their ideas about deterrence was in itself "an important component of its deterrent doctrine."[61]

But despite the advantages of this "calculated ambiguity" in Chinese strategic doctrine, there were also dangers. The view that a power is irresponsible and perhaps irrational may encourage the formation of an alliance against it and the attempt to isolate that state in the world community. A similar impression may also lead to a preemptive strike against the state's nuclear facilities to prevent the emergence of a nuclear power that seems to endanger world peace. Fear of both of these possibilities may largely account for the major shift in Chinese foreign policy in 1971. The threat of a Soviet attack and of Soviet-American collusion seems to have caused a reassessment of Chinese policy to lessen the dangers of both. Those who supported the new policy in Peking may have hoped that a partial rapprochement with the United States would enhance China's national security by discouraging a Russian attack and breaking the alliance that was developing between the two superpowers. Chinese tactics seem to have had some limited success.

During the late 1970s the Chinese also seemed to be moving away from the "calculated ambiguity" of previous policy, or at least technological developments suggested a somewhat clearer policy. The deployment of intermediate-ranger bombers, medium-range ballistic missiles (MRBMs), and intermediate-range ballistic missiles (IRBMs) in the early 1970s was followed by the development (and probable deployment) of a limited-range, multistage ICBM as well as a submarine-launched ballistic missile (SLBM) capability. The Chinese continued during the late 1970s to demean the role of nuclear weapons in combat while plowing considerable resources into developing an invulnerable sea-launched nuclear system. The overall nuclear program has, however, remained limited. Despite the continuing lack of a public nuclear debate, these technological developments suggest that China is content to depend, for the present at least, on a minimum deterrent posture. Chinese officials have indicated that nuclear weapons are needed to offset the superpowers' monopoly but that there was no requirement to match the United States or the Soviet Union in the numbers of nuclear armaments. Having achieved an invulnerable retaliatory capacity given the limited resources available, Chinese leaders appear to wish to balance expenditure on nuclear weapons with that of conventional forces (and that of defense modernization with the other modernization programs).

China's emergence from isolation to play a greater role in the international community and her concern to improve her strategic vulnerability may, as time goes on, force her to define her views on deterrence more clearly in public than she has done in the past. As Chinese nuclear forces develop further, the thinking behind the development is likely to become more obvious to the outside world. The strengthening of the Chinese force may also perhaps lead to an increased confidence in Peking, which will encourage Chinese leaders to enter into the strategic debate with the other two superpowers as a means of achieving security. In the past the doctrine of "strategic ambiguity" was largely a product of China's weakness.

The speed of the long march to modernization is likely to be dictated to a certain extent by the outcome of the debate between those who seek outside (at present, Western) military technology and those who seek a more self-sufficient future for China. Despite the great interest that has been shown in foreign equipment and weapons, as yet the Chinese appear to be relying upon largely homegrown solutions to their security problems. It may be that this will change, especially if China adopts a true strategic partnership with the Western powers. There is little doubt that the PLA has major deficiencies across the board, and the purchase of sophisticated Western military hardware would help to improve its operational capabilities. The limitations of the Chinese economy and the political priorities established limit how much Western technology can be afforded, however, even within the present framework of Chinese relations with the West. The contemporary difficulties in Sino-American relations and traditional Chinese views about the transience of alliances could also result in a strategic shift, causing China to go it alone—maintaining its distance as it did in the past from both the Soviet Union and the United States. Equally, the pragmatism of Chinese policy could see China, under changing strategic and political circumstances, once more "leaning to one side" toward the Soviet Union. Like other aspects of policy the future of Chinese security policy will be determined to a large extent by domestic political circumstances. Changes in these circumstances could result in new strategic reassessments and a domestic and international role for the PLA very different from that envisaged at present. Additionally, increased technological sophistication as well as greater contacts with other states may in time not only lead to a more public debate in China but also to Chinese participation in the strategic dialogue with other nuclear powers.[62]

## *Conclusion*

Chinese defense policy for some time to come is likely to be based as it is today on the twin pillars of deterrence provided by the concept of "people's war under modern conditions" and nuclear armaments.[63] "People's war under modern conditions" attempts, through the mass mobilization of the nation's population and the adoption of Mao's three-stage strategy to deter or repel any conventional invasion by a foreign power. Nuclear deterrence, on the other hand, is designed to deter a strategic attack on China through a threat to use the most sophisticated weapons of modern technology. There is a sense in which these two themes represent the product of the continuing struggle between revolutionary principles and modernization or professionalism, which, it has been argued, have pervaded Chinese defense policy since 1949. Although Chinese defense policy has often in the past been the product of a compromise between these two influences, the death of Mao and the demise of the Gang of Four have significantly tilted the balance in favor of professionalism. As we have argued, however, despite the greater sensitivity toward the PLA's problems within Chinese leadership circles, military modernization is not being given undue priority. Defense improvements are seen as taking place alongside the other three modernization programs (the Four Modernizations Program also involves agriculture, industry, and science and technology) with priority being given to the overall development of the Chinese economy. Thus, while the present leaders remain in power, efforts will be made to improve the professional training of the PLA, its equipment, and its weapons, but the process will be gradual.

## NOTES

1. Coral Bell, "China: The Communists and the World," in F. S. Northedge, ed., *The Foreign Policy of the Powers* (London: Faber and Faber, 1974), 120.
2. Robert A. Scalapino, "The Cultural Revolution and Chinese Foreign Policy," *Current Scene: Developments in Mainland China* 6, no. 13 (1 August 1968).
3. Sun Yat-sen, *San Miu Chu I* (Three People's Principles) (Taipei, Taiwan: Cheng Chung, 1954), 6; quoted in Liu, *China as a Nuclear Power in World Politics* (London: Macmillan, 1972), 14.
4. Arthur Huck, *The Security of China* (London: Chatto and Windus, 1970), 20.
5. For a discussion of similar statements by Chiang Kai-shek and Mao Tse-tung, see Huck, ibid., 21–22.
6. See Liu, *China as a Nuclear Power*, 11.
7. *People's Daily*, 31 March 1964.
8. Franz Michael, "A Design for Aggression," *Problems of Communism* (Jan.–April 1971): 63.
9. Peter Van Ness, "Mao Tse-tung and Revolutionary 'Self-Reliance,'" *Problems of Communism* (Jan.–April 1971): 69. Much of the argument that follows in this section owes a great debt to this article by Professor Van Ness.
10. Ibid.
11. Allen S. Whiting, *China Crosses the Yalu* (Stanford, Cal.: Stanford University Press, 1968).
12. See, for example, Neville Maxell, *India's China War* (New York: Pantheon, 1970); and Allen S. Whiting, "The Use of Forces in Foreign Policy by the People's Republic of China," *Annals of the American Academy of Political and Social Sciences* 402 (July 1972); and *The Chinese Calculus of Deterrence* (Ann Arbor: University of Michigan Press, 1975).
13. For a development of the argument see Whiting, "The Use of Forces," 58–59.
14. See Harold C. Hinton, "Conflict on the Ussuri: A Clash of Nationalisms," *Problems of Communism* (Jan.–April 1971); and Van Ness, "Mao Tse-tung and Revolutionary 'Self-Reliance.'"
15. Van Ness, ibid., 70.
16. See *Strategic Survey 1979* (London: International Institute for Strategic Studies, 1980), 59.
17. Van Ness, "Mao Tse-tung and Revolutionary 'Self Reliance,'" 70.
18. Michael, "A Design for Assertion," 63.
19. Daniel Lovelace, *"People's War" and Chinese Foreign Policy: Thailand as a Case Study of Overt Insurgent Support*, unpublished Ph.D. dissertation, Claremont Graduate School, 1970. See 215–18, quoted by Van Ness, "Mao Tse-tung and Revolutionary 'Self-Reliance,'"

71; see also Daniel Lovelace, *China and People's War in Thailand, 1964–69,* China Research Monograph no. 8 (Berkeley, California: University of California Press, 1971).

20. Quoted in John Gittings, *The Role of the Chinese Army* (London: Oxford University Press, 1967), 32.

21. See ibid., 40.

22. Whiting, "The Use of Forces," 63–64.

23. Besides the invasion of Tibet and the preparations to invade Taiwan, the PLA was also involved in this period in "bandit suppression." See Gittings, *The Role of the Chinese Army,* 32–37.

24. See *Strategic Survey 1981–2* (London: International Institute for Strategic Studies, 1982), 101–4.

25. Gittings, *The Role of the Chinese Army,* 182. See also Stephen A. Sims, "The New Role of the Military?" *Problems of Communism* (Nov.–Dec. 1967): 28.

26. As Gittings points out, the campaign was designed to "publicize throughout the country the techniques and activities of political education and control which had been practiced in the PLA since 1960." See Gittings, *The Role of the Chinese Army,* 256. The PLA was also used in important administrative and peacekeeping roles during the GPCR.

27. See Ellis Joffe and Gerald Segal, "The Chinese Army and Professionalism," *Problems of Communism* 27 (Nov.–Dec. 1978).

28. Gittings, *The Role of The Chinese Army,* 25.

29. The conference specified that China should have a "unified army," which should be modernized, and that an air force and navy should be established. See Gittings, ibid., 119.

30. *People's Daily,* 1 January 1952; quoted in Gittings, ibid., 25.

31. See Harry Harding, Jr., "The Making of Chinese Military Policy," in W. W. Whitson, ed., *The Military and Political Power in China in the 1970s* (New York: Praeger, 1972), 375.

32. See David A. Charles, "The Dismissal of Marshall P'eng Teh-huai," *China Quarterly* (Oct.–Dec. 1961).

33. See Maury Lisann, "Moscow and the Chinese Power Struggle," *Problems of Communism* (Nov.–Dec. 1969): 34.

34. See Harding, "The Making of Chinese Military Policy," 377.

35. Ibid., 379.

36. See Jurgen Domes, "New Course in Chinese Domestic Politics: The Anatomy of Readjustment," *Asian Survey* 13, no. 7 (1973); and Parris H. Chang, "Regional Military Power: The Aftermath of the Cultural Revolution," *Asian Survey* 12, no. 12 (December 1972).

37. See Henry S. Bradsher, "China: The Radical Offensive," *Asian Survey,* 12, no. 11 (November 1973): 1004.

38. See Joffe and Segal, "The Chinese Army and Professionalism."

39. Ibid.

40. Ibid.

41. "Statement by the Spokesman of the Chinese Government—A Comment on the Soviet Government's Statement of 21 August," *People of the World, Unite, for the Complete, Thorough, Total and Resolute Prohibition and Destruction of Nuclear Weapons* (Peking: Foreign Languages Press, 1963), 30; quoted in Liu, *China as a Nuclear Power,* 34.

42. Gittings, *The Role of the Chinese Army,* 119–20; and *Peking Review,* 8 May 1964.

43. See Gittings, *The Role of the Chinese Army,* 232.

44. See Liu, *China as a Nuclear Power,* 34.

45. See Huck, *The Security of China,* 65.

46. B. W. Augenstein, "The Chinese and French Program for the Development of National Nuclear Forces," *Orbis* 11, no. 3 (1967).

47. See *Strategic Survey 1981–82,* 103.

48. William R. Heaton, Jr., "The Defense Policy of the People's Republic of China," in Douglas J. Murray and Paul R. Viotti, eds., *The Defense Policies of Nations: A Comparative Study* (Baltimore: The Johns Hopkins Press, 1982), 436.

49. See *Strategic Survey 1981–82*, Chinese section.

50. Ralph L. Powell, "Maoist Military Doctrine," *Asian Survey* 8, no. 4 (April 1968): 239–67.

51. Ibid., 243.

52. Huck, *The Security of China*, 61.

53. Ibid.

54. Jack H. Harris, "Enduring Chinese Dimensions in Peking's Military Policy and Doctrine," *Issues and Studies* (July 1979); and William R. Heaton, Jr., "The Defense Policy of the People's Republic of China."

55. Xu Xiangqian, "Heighten Our Vigilance and Get Prepared to Fight a War," *Peking Review*, 11 August 1978.

56. G. Segal, "The Soviet Threat at China's Gates," unpublished paper.

57. Xu Xiangqian, "Heighten Our Vigilance."

58. Ibid.

59. See M. Halperin, "China's Strategic Outlook," in Alastair Buchan, ed., *China and the Peace of Asia* (London: Chatto and Windus, 1965), 101.

60. This is the term used by Harry Gelber in "Nuclear Weapons and Chinese Policy," *Adelphi Paper* no. 99 (London: International Institute for Strategic Studies 1973), 21. For an alternating view see Gerald Segal, "China's Security Debates," *Survival* (March–April 1982). See also G. Segal, *Defending China* (London: Oxford University Press, 1985).

61. Gelber, "Nuclear Weapons and Chinese Policy," 22.

62. Ibid.

63. See Heaton, "The Defense Policy of The People's Republic of China," 425–26.

# 5

# *British Defense Policy*

JOHN BAYLIS

British defense policy since 1945 has been characterized largely by a process of continual adaptation to the changing circumstances and problems caused by the devolution of empire. Fundamental changes in Britain's foreign policy throughout the postwar period have led inevitably to important changes in the defense sector. The military instrument that for so long had been predicated on sustaining the global responsibilities inherent in great-power status have increasingly had to be modified and adjusted to suit Britain's declining diplomatic role in the world.

A superficial examination of the relationship between foreign and defense policies reveals that in general the two related areas have to a large extent been in line. Britain's diplomatic posture for much of the postwar period, despite the gradual process of decolonization, was grounded on the traditional world role, and defense policy was directed toward providing the worldwide capability to perform the tasks inherent in such a role. To a large extent also the diplomatic withdrawal from empire has been paralleled, and in most cases symbolized, by the gradual contraction of the military effort toward a concentration on Europe.

A much closer examination of the relationship, however, reveals that in many other respects foreign and defense policies have never in fact been very closely related.[1] A more detailed analysis of the process of decolonization in particular suggests that defense arrangements were not always adequately related to the timetable and consequences of withdrawal. It would seem that very often decisions about defense requirements were taken by the Ministry of Defence in apparent disregard of the policies of withdrawal that were being taken by the Foreign and Colonial offices.[2] The failure of the departments concerned to coordinate foreign and defense policies in this important field meant that instead of an overall, integrated course of action, several independent and often contradictory lines of policy were often pursued.

The result of this lack of overall political direction and coordination has been that defense policy throughout the period has been in a state of flux as attempts have been made to meet changing international as well as domestic circumstances.[3] In one sense the problem has been partly psychological. Decision makers responsible for the formulation of policy in both the foreign and defense fields have been constantly faced with the problem of reconciling past images with the harsh realities of the moment. The clash between the traditional image of Britain as a great world power and the increasing reminders of her reduced diplomatic, military, and economic capabilities produced a psychological environment during the period after 1945 that made it difficult for those responsible for making important foreign and defense policy decisions to constantly modify and readjust policy.[4]

This process of constant adaptation and readjustment has taken place in most of the more important areas of policy. Continuous adjustments in emphasis, for example, have taken place between the three overlapping circles of British external relations: the Commonwealth, the United States, and Europe. In consequence, the military has constantly adapted to the changing emphasis on these different areas and the disparate variety of operations this has necessitated. Adjustment in emphasis has also been made between commitments and capabilities, between conventional and nuclear forces, between ideas concerning total war and limited war, and even between service autonomy and centralization within the machinery responsible for defense policy formulation. Because these issues and the process of adjustment they reflect characterize much of British defense policy since 1945 it may be useful to focus attention on each of the questions in turn. In so doing it is hoped to provide a coherent picture of the essential features of the evolution and determinants of policy in the postwar period.

## *Readjustment of Roles*

The immediate postwar period was essentially one of transition and uncertainty in defense policy as planners attempted to take stock of the changed external and domestic circumstances in the postwar world. Attlee saw the 1946 defense white paper as a stop-gap, and Athony Eden claimed that it was not so much a "statement relating to defence but a progress report on demobilization."[5] This cautious and tentative approach of the government toward defense in these years was reflected in the adoption of a ten-year rule in 1946, which, like the ten-year rule of 1919, was a directive to the military to plan on the official assumption that there would be no major war for a decade.[6] In this uncertain environment, with no identifiable enemy as yet to plan against, defense planners searched for a viable security system for the future. With the mood of disengagement in the United States, some form of Commonwealth defense arrangement seemed to many to be the most fruitful.[7] On the basis of the decision of the Imperial Conference of 1926, the general defense of the empire, in the form of the defense of the empire's lines of communication, was the responsibility of the United Kingdom, assisted by such forces as the dominions could provide. Although by 1945 it was recognized that this scheme was obsolete, it was hoped that the system of informal cooperation would continue with greater sharing of the burdens of Commonwealth defense. In 1946 and 1947 this idea was accepted when the principle of allocating zones of responsibility within the Commonwealth was adopted by a number of members. A system of regional association was established, and overall cooperation obtained by the creation of a system of Commonwealth liaison officers.

Although this loose arrangement was subsequently formalized to a certain extent with the ANZAM agreement of 1949, when the governments of the United Kingdom, Australia, and New Zealand agreed to coordinate defense planning, the Commonwealth defense idea was not to prove the long-term solution to British security problems many had hoped that it would. The hopes of the chiefs of staff in 1945–46 that India would become the center of the regional Commonwealth grouping had to be abandoned with the loosening of ties between the two countries after Indian independence in 1947. The movement of Australia and New Zealand toward increasing dependence on the United States for their security following the ANZUS treaty of 1951 also helped to create a feeling of disillusionment in Whitehall with Commonwealth defense arrangements.

Events in Europe proved critical in directing the attention of defense planners away from the idea of Commonwealth defense. The coup in Czechoslovakia and the Berlin crisis in 1948 finally convinced the Labour government that the Soviet Union was the potential enemy and that the greatest danger to British security lay in Europe. This conviction led Bevin to try to "entangle" the United States in the defense of Europe and reestablish the "special relationship" between Britain and America that had been built up during the Second World War.[8]

Relations with the United States following the end of the war had deteriorated with the end of Lend-lease, the passing of the McMahon Act, which virtually halted Anglo-American nuclear collaboration, and disagreements over Palestine. The growing American perception of Soviet intransigence and the gradual emergence of the cold war, however, led to a reversal of the process of demobilization and retreat to isolationism in the United States.

The movement away from the Commonwealth defense idea therefore led increasingly throughout the late 1940s toward alliance defense arrangements predicated largely on American assistance. The renewal of the bilateral Anglo-American "special relationship" in the defense field after the relapse of 1945–46 had begun (tentatively) in the latter part of 1946 when there was a continuation of staff contacts and exchange of officers. The exchange of air force officers in particular helped facilitate the dispatch of two groups of American B-29s to Britain in July 1948 as part of the Western response to the Berlin crisis.

In February 1947, after Britain's famous warning that the state of the economy made it impossible to continue to assist the Greek government in the civil war against its Communist opponents, the United States stepped into the breach under the banner of the Truman Doctrine. Britain was also largely responsible for tying the United States much closer to Europe in this period, firstly through economic aid under the Marshall Plan and secondly, following the Western European Brussels Pact of 1948, through the North Atlantic Treaty Organization, established on 4 April 1949, which represented a major triumph in postwar British diplomacy.

The most productive part of the evolving relationship with the United States in defense matters was, however, in the nuclear field. After Anglo-American nuclear cooperation ceased following the McMahon Act of 1946, Britain continually pressed the United States to resume collaboration and amend the act. Up until the testing of its first atomic bomb, Britain was largely unsuccessful in restoring wartime cooperation. With the Monte Bello test on 3 October 1952, however, when it was evident that Britain had joined the nuclear club, the hallowed doors of nuclear cooperation gradually opened up. The gradual easing of restrictions began with the *modus vivendi* of 8 January 1948 and was followed by the Atomic Energy Act of 1954, which permitted the sharing of the data on the external characteristics of nuclear weapons in terms of size, weight, shape, yield, and effect.

The climax to Britain's postwar attempts to share nuclear information with the United States came in 1958 with the amendments to the Atomic Energy Act of 1954. Following the serious break in Anglo-American relations over Suez in 1956, Macmillan, on becoming prime minister, moved swiftly to restore the harmony of the "special relationship," which had been ruptured in the Anglo-French invasion of Egypt. The most important outcome of this British diplomatic offensive was the amendments to the 1954 Act whereby Britain was given even more preferential treatment in the nuclear field than it had achieved hitherto. The new amendment permitted the exchange of information about the design and production of nuclear warheads and the transfer of fissile materials to countries that had made substantial progress in the development of atomic weapons. As Britain was the only ally to qualify under this stipulation, it was the only beneficiary from the amendment.

Suez was also ironically a milestone in the return to collaboration in the field of delivery systems. Macmillan's Bermuda meeting with President Eisenhower in March 1957 resulted in the installation in Britain of sixty Thor liquid-fueled intermediate-range ballistic missiles (IRBM) under a two-key system. At the time Britain's delivery system was based on the V bomber force, and the projected Blue Streak missile was largely independent of the United States. Following the launching of Sputnik by the Soviet Union on 4 October 1957 and the subsequent advances in the accuracy of Soviet and American ballistic missiles, however, the British delivery system became increasingly vulnerable. To overcome its weakness and also because of the system's escalating costs, Blue Streak was canceled in April 1960, and the decision was taken to rely more heavily on the Americans by buying Skybolt from them to augment the V bomber force.

The Anglo-American "special relationship" was temporarily jeopardized once again in the autumn of 1962 when the Americans decided to end research and development on the Skybolt missile allegedly because of the increasing vulnerability of aircraft on the ground to a surprise first strike by Soviet missiles. In December 1962 at the Nassau meeting, however, President Kennedy's offer of Polaris missiles did a great deal to improve and consolidate Anglo-American relations and help preserve in greatly modified form Britain's independent deterrent.

Despite improved Anglo-American relations in the diplomatic as well as in the defense fields after the Suez incident, the rupture in relations it caused led, although belatedly and reluctantly, to a reappraisal of Britain's role in the world and the gradual movement toward concentrating its political, economic, and military effort on Europe.

Throughout much of the postwar period, although defense planners had been reluctant to concentrate the defense effort in Europe, the Continent had nevertheless been a high priority in terms of Britain's commitments. In 1947 Britain had signed the Dunkirk Treaty with France against the possibility of a resurgent Germany, and in 1948, with the clear identification of the Soviet threat, the Brussels Pact was formed to establish a larger defense structure to facilitate Western European security.[9] This was followed by the North Atlantic Treaty of 1949, whereby Britain managed to achieve not only further European collaboration but also the long-sought-after American guarantee to defend Europe against Soviet invasion.

Symptomatic of Britain's ambivalent attitude toward Europe was the debate that continued throughout the late 1940s and early 1950s over whether Britain should adopt a continental or a "limited liability" strategy. The traditional penchant for a limited contribution to European defense was reinforced after the Korean War when the fear of an imminent Soviet attack on Western Europe had died down. In 1952 almost immediately after the Lisbon Conference, Britain took steps to water down the overoptimistic force levels accepted at that NATO meeting. The 1952 global strategy paper in particular argued that manpower targets could be considerably reduced because of the advent of tactical nuclear weapons.[10] European defense remained important, but the government was unwilling, with its worldwide commitments, to concentrate effort and resources in Europe.

The government was forced, however, to compromise its position over the European commitment a little in 1954 when the question of German rearmament and the European Defence Community threatened to split the Western alliance. On the one hand German rearmament was needed to meet the growing conventional deficiency between East and West, but on the other hand lingering French fears of a revisionist Germany had to be allayed. Through its proposals at Paris in 1954, Britain was able to secure German rearmament inside NATO and at the same time safeguard French sensitivity over the German issue by guaranteeing to keep a force of four army divisions and a tactical air force (or the equivalent) stationed in Europe until 1994.

Although this commitment was significant in that it represented a reversal of traditional British policy toward Europe, throughout the 1950s and early 1960s the relative stability engendered in Europe through the creation of NATO and the Warsaw Pact led to a decreasing emphasis in British defense policy on European affairs. There was a concentration instead on fulfilling the continuing tasks and commitments overseas in the Far East and Middle East. The neglect of Europe to a certain extent, symbolized by the borrowing of troops from NATO at times to alleviate the strain on forces east of the Suez, continued until the 1960s. Only then did financial issues and the dangerous imbalance between commitments and capabilities lead to a fundamental reassessment of defense priorities and the concentration on Europe. The new European orientation initiated by the first application to join the Common Market in 1961 was taken further in the defense field with the limited retrenchment announced following the 1966 defense review and came to fruition finally with the decision to withdraw forces from the Far East and Persian Gulf after 1971.[11]

## *Military Operations*

This gradual and uncertain process of retreat from empire and the constant search throughout the postwar period for a new role has had a significant impact on the kind of military operations the armed forces were called upon to perform.

At the end of the Second World War, the armed forces were immediately involved in numerous military commitments throughout the world. Quite apart from the inescapable need for armies of occupation in Europe—in Germany, Austria, and Venezia Giulia—there was also a need to occupy and control various colonies, both her own and those of other powers. Occupation armies were needed in particular in ex-Italian colonies in Africa and also in many parts of the Far East, which had until recently been occupied by the Japanese. Also, quite apart from British territories such as Malaya, Singapore, Burma, and Hong Kong, both

British and Indian forces assisted in the reoccupation of the former Dutch and French colonies in the East Indies and Indochina. These latter duties soon brought them into direct conflict with nationalist forces until the old colonial powers were able to return.

Although the liquidation of occupation responsibilities was not finally completed until the mid-1950s, the focus of attention of military planners from 1947 to 1948 onward centered increasingly on the two main, and often competing, strands of British foreign policy: the European and the overseas roles. In Europe after the successful show of force against the Yugoslavs over Trieste in 1945 and the campaign against the Communist forces in Greece in 1946–47, the armed forces settled down to a peacetime role of operating within an alliance—first the Western European Union (WEU) and later NATO—to deter Soviet aggression in Western Europe.

The kinds of skills and equipment required for a role of this sort were totally different from those needed for operations in overseas territories. As the British Army of the Rhine (BAOR) trained for large-scale conventional and later nuclear, chemical, and biological operations in Europe, the gradual and often reluctant withdrawal from empire increasingly necessitated policing, antiterrorist, and anti-insurgent operations in many different environments and terrains as the tide of nationalist sentiment demanding independence grew. As one writer rather elaborately puts it: "Just as the marches of neglected, underpaid redcoats and their thundering volleys had once marked the surge of British expansion, so the soldier in Khaki drill or battledress guarded the hesitant retreat from Empire. While the politicians wavered, while authority crumbled, the soldier tried to keep order."[12] Although this view rather underestimates the difficult tasks that statesmen faced and the often intractable problems with which they grappled, the fact remains that the armed forces were often left to "hold the ring" against indigenous rioting and terrorist activity as politicians reluctantly worked out their timetables of withdrawal. This pattern of events was repeated in the Middle East, in Africa, in the Carribean, and in Asia and the Far East.

In the Middle East, the armed forces conducted operations against terrorist groups in Palestine, in Egypt, and in Cyprus in the late forties and early fifties as British governments, despite the declining importance of the bases in these areas, attempted to find political solutions that would guarantee a continuing presence; only to find that at the end of the day they were compelled to withdraw. The same was true of British possessions in Africa, Asia, and the Carribean, where between 1947 and 1967 the military was involved in maintaining order at one time or another in British Guiana, British Honduras, Kenya, Aden, the Gold Coast, Hong Kong, Jamaica, the Cameroons, Zanzibar, Borneo, Tanganyika, Uganda, and Mauritius. In the Far East also British troops successfully fought a protracted counterinsurgency campaign in Malaya against Communist guerrillas between 1948 and 1960, which enabled the colony to progress smoothly to independence with a relatively stable democratic regime.[13] A similar operation with equal success was fought in Borneo between 1963 and 1966 when Indonesia's President Sukarno attempted to prevent the integration of Sarawak and Sabah into the newly formed Malaysian Federation through his policy of confrontation.[14]

More recently British troops have also become involved in counterterrorist activities much nearer home against the IRA in Northern Ireland. Despite some similarities with counterinsurgent campaigns in other countries, operations in the British Isles, in a predominantly urban environment, have required new techniques and created new problems that the army has had to solve. Once again the armed forces have been asked to deal with the bombings, snipings, and assassinations of the terrorist campaign while the politicians grapple with the complex and confusing political issues in an attempt to find a satisfactory solution to Ulster's problem.

In contrast to these irregular campaigns the armed forces have also been engaged in others, requiring vastly different expertise and training. In June 1950, for example, British forces became involved in fighting a conventional war against Communist forces following the invasion of South Korea by forces from the north. For defense planners in Britain the campaign against North Korea in support of the south was important from two points of view. On the one hand, and most obviously, the Commonwealth Brigade was important in helping the United States and the United Nations to prevent Communist aggression and expansion in Korea and thus indirectly in the Far East as a whole—aggression that, if allowed to spread, would, it was thought, threaten British interests in the area.[15] It was more important, however, from another British perspective. To many in Western Europe the invasion of South Korea was seen merely as a feint, an act of deception by Stalin, to distract the attention of the Western world away from Europe prior to a concerted Soviet invasion. In those early days of NATO and American involvement in Europe that Ernest Bevin had worked so hard to achieve, the Attlee government considered that it was vital for political reasons to demonstrate British, and therefore Western European, support for the United States in the Far East to ensure continued American military support in Europe against the perceived Soviet threat.

In the majority of cases, British military operations, whether of a policing, antiterrorist, counterinsurgent, or conventional variety were largely successful in achieving the main political objectives laid down. The relative utility of military power as an instrument of state policy in the postwar period has, however, derived not only from the successful use of military force in various circumstances but also often from the threat of the perceived potential that British forces symbolized. This was particularly the case in the Kuwait operation in 1961, when the landing of British forces there forestalled a threatened annexation by neighboring Iraq. One study of the operation points out that with all its inadequacies from the British side and the Iraqi numerical superiority and largely superior equipment, even taking into account the more nebulous question of morale, training, and leadership, the British would have been very hard-pressed to meet a concerted invasion by Iraq had it taken place.[16] As it was, despite the deficiencies of the military operation, Kassem did not invade largely perhaps because of the commitment the British government had made and the potential the forces on the ground represented.

Despite the relative inadequacies, particularly of equipment, highlighted by the Kuwait operation, the forces that existed did enable the government to respond, even if partly fortuitously, to a perceived problem in a relatively short period of time. This was not the case, however, during the Suez crisis in 1956. Despite the lip service that had been paid to the concept of a highly mobile strategic reserve in the white papers from 1954 onward, the overriding importance of the deterrent and the limited resources available precluded any major improvement in conventional capability. Consequently, when the government was faced with a situation over the nationalization of the Suez Canal in 1956, because of the deficiencies that existed it was unable to respond with the kind of swift and mobile operation that would have been necessary to achieve the desired results.[17] Although the crisis broke on 26 July, the military preparations were not complete until mid-September, and even then after the initial bombardment of Egyptian airfields on 31 October, the Anglo-French assault on Port Said did not begin until 5 November because of the long sea voyage from Malta.

Suez, in many respects, represented a failure of British military power to deter actions that were regarded as hostile to the nation's vital interests. Twenty-six years later much the same thing was to occur (with a different outcome) when the Argentinians invaded the British territories of South Georgia and the Falkland Islands. The defense review of 1981, which announced naval cuts, seemed to suggest that Britain was taking further steps in the abandonment of its worldwide role. Together with domestic difficulties in Argentina, the projected cutback convinced the Argentine junta that the time was ripe for a national war to recover the Malvinas. The junta, however, underestimated Britain's political resolve and continuing military capability to operate in defense of its interests far from its own shores. As the post-Falklands white paper commented:

> In the space of seven weeks a task force of 28,000 men and over 100 ships had been assembled, sailed 8,000 miles, effectively neutralised the Argentine navy and fought off persistent and courageous attacks from combat aircraft which outnumbered its own by more than six to one. This in itself was no mean feat, but the task force then put ashore

> 10,000 men on a hostile coast while under threat of heavy air attack; fought several pitched battles against an entrenched and well supplied enemy who at all times outnumbered our forces; and brought them to surrender within three and a half weeks.[18]

With the Argentine surrender on the 15 June, British forces had demonstrated the quality and flexibility of British military manpower, conventional weapons, and equipment. Question marks remained, however, over certain aspects of the campaign and the conclusions of the defense review of 1981.

The Suez operation and the Falklands conflict symbolized the kinds of problems that military planners have faced in the postwar years. Successive governments throughout the period, largely because of their assumptions of great-power status, have continually called upon the military to perform a great variety of tasks around the world and at the same time have reduced the manpower of the armed forces and drastically cut the expenditures available for the variety of weapons necessary for such operations. Under these circumstances the military forces have been forced to adjust continually to the new circumstances and improvise as best they can. This was true in both the Suez crisis and the Falklands war, as well as in many of the other campaigns undertaken since 1945. The failures when they have occurred, have usually reflected a lack of coherent and clear political direction rather than of operational skills and military performance. This was true of Suez and perhaps of the Aden operations as well.

## *Commitments and Capabilities*

The constant adjustment and search for a new role and the pressures on the services performing many different operations have to a large extent been a consequence of the imbalance that has existed for much of the postwar period between commitments and capabilities. In 1947 the decision was taken to give India independence, thus in part relieving Britain of one of its biggest commitments. The problem for British defense planners, which does not seem to have been immediately recognized, was that it also deprived it of Indian military manpower, which had been traditionally used to garrison the empire. Despite the loss of the cornerstone of the empire, the remaining areas of British sovereignty, various residual obligations, and perhaps most importantly "the habit of influence," all helped to sustain the impression that Britain remained a great power in Asia. At the same time conditions in Europe required active participation in an alliance with the maintenance of substantial conventional forces permanently on the ground.

The attempts to fulfill this dual defense policy between 1947 and 1967 put increasing strain on the defense establishment to perform numerous commitments with decreasing manpower and financial resources. Following the loss of Indian manpower, various decisions were made in 1956 and 1957 to cut service manpower, largely for economic reasons. Eden announced the reduction of forces from 800,000 to 700,000 men in 1956 and planned for a reduction to about 450,000 by 1960 or 1961. His plans were cut short by the Suez crisis; but in the reappraisal of military policy that followed, Duncan Sandys, the new defense minister, planned for an abolition of conscription by 1962 and the reduction of service manpower from 690,000 to 375,000 by that year.[20] The services were therefore being asked to continue performing traditional commitments during the early 1960s with half of the manpower available in 1956.

There were also the problems caused by rising prices and technological advances, which were made considerably more difficult by the gradual reduction in defense spending. Following the massive rearmament program initiated by the Labor government in 1950, the plan to spend 10 percent of GNP on defense was gradually "stretched out" by the Conservative government such that it was finally reduced to 6.6 percent when the Conservatives left office in 1964. At the same time the cost and complexity of weapons systems, particularly those required to retain a global military capability and nuclear delivery system, began to escalate rapidly. The aircraft carriers that in the 1930s had cost £3 million were costing £60 million by the mid-1960s, and the cost of army equipment increased fourfold in the twenty years after the war. As Michael Howard rightly points out, "in order to continue to carry out their traditional tasks, the services were compelled year by year to demand bigger appropriations from a Ministry of Defence which year by year had a smaller proportion of the national income from which to meet their requirements."[21]

In the years after the Kuwait operation in 1961 as commitments east of Suez were increasingly tested in Southeast Asia, south Arabia, and East Africa,* the material and manpower difficulties created new problems. The 1966 defense review emphasized the overstretch of British forces but did little either to reduce the commitments or increase the capabilities of British military forces by providing the manpower, mobility, and resources that were necessary. The difficult political decisions required to improve the balance were eventually brought about largely by financial crisis rather than rational strategic planning. It was felt that the continuing economic problems required major cuts in government spending. In order to make cuts in social expenditure more palatable to various sections of the Labour party, the government felt it necessary to introduce retrenchments in the defense field as well.[22] In 1967 and 1968 therefore, faced with a rapidly deteriorating economic position, the Wilson government

*The operations included Brunei, Borneo, Aden, and the East African mutinies.

finally decided to withdraw forces from east of Suez, thereby cutting commitments and bringing them more into line with the resources available. As one commentator put it, the right decision had been made but largely for the wrong reasons.[23]

The adoption of a more "appropriate" defense policy after the defense reviews of the late 1960s was not, however, the end of the story. Faced with continuing economic difficulties the Labour government of 1974 undertook what it described as "the most extensive and thorough review of our system of defence ever undertaken by a British Government in peacetime."[24] In an attempt to cut defense spending from the existing level of 5.5 percent of the GNP to 4.5 percent over the next ten years a further geographical contraction of Britain's defense effort into, and even within, the NATO alliance took place alongside cuts in equipment and force levels. Britain's defense efforts were now to be concentrated on four main areas: the strategic nuclear deterrent; defense of the home base; a contribution to the defense of the central front; and defense of the eastern Atlantic. Only residual forces were now to be retained for defense responsibilities that remained outside Europe.

Despite further cuts in defense in 1975 and 1976 the priorities established in the 1974 defense review were to remain in force into the 1980s.[25] By then, however, the effects of defense inflation and rapid technological change required a new attempt to balance commitments and capabilities, even though Labour and Conservative governments committed themselves to implement the NATO guidance of 1978 to increase defense spending in real terms by 3 percent per annum. The controversial 1981 defense review entitled "The Way Forward" reflected the government's judgment that improvements were necessary in the strategic nuclear deterrent and home defense.[26] Despite a decision to streamline BAOR, the review also reaffirmed Britain's commitment to station fifty-five thousand men on the continent to help defend the central front. The main change announced was a reduction in the surface fleet component of the defense of the eastern Atlantic. Altogether the numbers of destroyers and frigates were to be reduced from fifty-nine to forty-two (including eight withdrawn to stand-by duty).

To many observers, especially those from the naval lobby, such a cut in general-purpose naval forces would reduce significantly Britain's capability to meet not only the Soviet threat in the eastern Atlantic but also unexpected threats to commitments outside Europe. It was therefore not surprising that the Falklands war in May–June 1982 should bring accusations of strategic myopia. The conflict was not only unexpected, but it was outside Europe and relied on the very service that Defense Secretary Nott was about to cut. For many therefore the war was seen as "a lesson in the nick of time, before the senior service was rendered incapable of coping with such eventualities." In the review of the Falklands war, which followed in December 1982, however, the government stuck largely to the priorities established in the 1981 review.[27] More stress was given to capabilities to operate outside Europe, but the commitment to the central front and the threat from the Soviet Union remained the primary focus of the armed forces. Despite the logic of the government's position, the dilemma of balancing commitments and capabilities remained very much the order of the day.

## *Nuclear and Conventional Forces*

This asymmetry between commitments and capabilities has been, to a large extent, due to the reluctance of succeeding governments to provide the necessary conventional forces or to cut commitments quickly enough. Here again, there has been a constant process of adjustment in emphasis between nuclear weapons systems and conventional forces.[28] With the watering down and stretching out of the Labour government's rearmament program by Prime Minister Churchill in 1951, emphasis in British defense policy increasingly became focused upon nuclear weapons. For most of the period from the global strategy paper of 1952 to the Chiefs of Staff paper of 1962 the policy of nuclear deterrence was stressed in the defense white papers. The most explicit statement of the nuclear deterrent doctrine came in 1957 with the Sandys white paper.[29] Following the humiliation of Suez, even when the United States had actively opposed British policies, it was felt that, both for political and military reasons, emphasis should be placed on building up an independent nuclear deterrent. The white paper announced that for the next five years Britain was to upgrade the priority accorded to the nuclear deterrent. Thermonuclear weapons would be added to the existing atomic armory, and the V bombers would be supplemented by ballistic rockets as they became available.

The significance of this was particularly important in relation to the provision of conventional forces. As Phillip Darby has put it, deterrence not only held the center of the strategic stage, but its shadows extended into the wings.[30] Despite the announcement in white paper after white paper that the government intended to improve conventional equipment, declaratory policy never really became actual policy, and resources were concentrated mainly on the nuclear field. Even after the increasing emphasis on providing limited-war forces for likely threats east of Suez contained in the 1962 Chiefs of Staff paper, limited improvements in mobility were stifled by the decision of the Labour government to impose a defense spending ceiling in 1964 of £2 billion. At the same time no attempts were made to cut commitments. The services, it would seem, were being asked to perform the same roles as before at less cost on the assumption that the Labour government could meet foreign and defense policy issues more skillfully than their predecessors.

The attempt by the Labour government between 1964 and 1968 to bring the nation's military capabilities into balance with its economic resources and foreign policy objectives, however, had its effects on the place and priority of nuclear weapons within overall defense policy. While in opposition the Labour party had by 1964 become rather belatedly critical of the emphasis the Conservative government was placing on nuclear weapons.[31] Once in power, however, the Labour party's policies were not as drastic as many had feared. The independent nature of the deterrent was played down, but the decision was taken to continue with the nuclear force despite the feeling of many in the party that it should be abandoned.

The most immediate problem facing the Labour government when it came to power was the question of the multilateral force (MLF) proposed by the United States to solve the issue of nuclear sharing in NATO. Prime Minister Harold Wilson's response, seemingly an attempt to sink the MLF proposal, was to put forward his own idea of an Atlantic nuclear force (ANF), whereby Britain's nuclear forces would be integrated into an alliance framework.[32] Although the proposal was of more political than military importance, it did set the tone of the Labour government's desire to play down national control of nuclear forces and emphasize alliance requirements. The difference between Labour and Conservative government policies toward nuclear weapons, however, was more verbal than substantive. In general the Labour government did not reverse the nuclear/conventional balance of British forces; rather, it attempted to integrate the two components more effectively in its overall strategic and foreign policy.

Minister of Defence Denis Healey's answer to the growing disparity between the conventional forces of NATO and those of the Warsaw Pact was the adoption of a strategy of "flexible escalation." This was an attempt to integrate and balance the deterrent and defensive elements of NATO strategy within a framework that would be politically and economically acceptable to member governments. The essence of the doctrine was that NATO should have sufficient conventional forces to deal with small-scale incursions or to hold a large-scale invasion for a few days while negotiations took place, after which time tactical nuclear weapons would have to be used. Healey's argument was that the enemy would be deterred because it would realize that NATO would have no alternative but to use tactical nuclear weapons at an early stage in an all-out conventional war. The initiation therefore of a large-scale attack in Europe posed the risk for the Soviet Union of escalation to strategic nuclear conflagration, a risk it would be unlikely to take.

During the 1960s and much of the 1970s the question of the balance between nuclear and conventional forces remained largely dormant. The preferential terms of the Nassau Agreement meant that little more than 1 percent of the defense budget annually had to be spent on maintaining a credible strategic deterrent force. By the late 1970s, however, the debate about the replacement for Polaris raised the issue once again.[33] The government's decision in July 1980 to purchase C4 Trident missiles from the United States and the later decision in March 1982 to purchase the even more expensive D5 missile created an outcry not just from nuclear disarmers but also from within the defense establishment itself. Many observers within the defense field argued that the expenditure of £8 to 10 billion or more on the new Trident force (alongside existing expenditure on Polaris through to the 1990s) would involve opportunity costs in other areas of defense. It was argued that in the middle and late 1980s when the bulk of the expenditure became due crucial choices would have to be made on important conventional weapons such as new tanks, new aircraft, and new frigates.[34] The firm commitment of the Conservative government of Prime Minister Thatcher to Trident seemed to many to be likely to lead to a deterioration in Britain's conventional forces in the 1990s despite the commitment to increases in defense spending.

## *Ideas about Total and Limited War*

The ongoing attempt to balance nuclear and conventional forces has been part of the attempt to integrate the general total-war* and limited-war notions into an overall coherent strategic framework. Up until the early 1960s strategic theory in Britain hinged on the assumption that the nations involved in war would devote their whole national effort to the object of winning the war on a global scale. British strategic thinking in this early postwar period was symbolized by the global strategy paper of 1952. The paper, which was prepared by the chiefs of staff at the prompting of Winston Churchill, contained three sections: one on deterrence, one on NATO, and the other on overseas responsibilities. Although the full rigor of the paper was never translated into policy, the primary point of the document relating to the importance of nuclear deterrence was immediately accepted and remained the basis of British defense policy for the remainder of the fifties. The general assumption behind strategic thinking was that the deterrent would be effective against threats of a cold war as well as a hot war. As the 1954 white paper pointed out, "As the deterrent continues to grow, it should have an increasing effect upon the Cold War by making less likely such adventures on the part of the Communist world as their aggression in Korea."[35]

Despite the continuation of the total-war theme throughout the 1950s, certain lines of policy emerged during the period, which were to harden

*The term *total war* is used here to refer to a conflict on a global scale in which the combatants employ all of their resources to achieve victory. *Total war* and *global war* are sometimes used synonymously. (See Darby, *British Defence Policy*, 76.)

into central prongs of British defense policy in the years after 1957 when the launching of the Soviet Sputnik made the strategy of massive retaliation less credible. The most important of the new developments initially was the plan for a central strategic reserve, which could be airlifted to overseas theaters when trouble broke out. Although this need for more mobile conventional forces remained a declared aim of the government from 1954, it was more important as a concept than as a policy until priority was given to building up transport command in the late fifties. The only service to take up the limited war theme with any enthusiasm in the midfifties, however, was the navy.[36] The navy's search for a new role after the broken-backed warfare concept* had been discredited caused it to settle by 1957 for a limited-war role. Although the Admiralty insisted that the navy still had an antisubmarine role in the event of general war, the limited-war priority became henceforward the basis of naval planning, and emphasis was placed on naval task forces.

The relative success of the Kuwait operation in 1961 crystallized ideas that had been prevalent for some time. A new Chiefs of Staff paper was produced in January 1962 that laid the emphasis much less on the total-warfare idea and more on the need to maintain the capability to counter limited threats that might emerge east of the Suez. From 1962 onward therefore, resources were increasingly channeled into providing greater mobility for the forces to perform small brushfire operations overseas. Transport aircraft became a higher priority in an attempt to make the airlifted strategic reserve concept a reality. More resources were allocated to the navy to develop its carriers as the central element in the seaborne task forces. This move toward upgrading the limited-war component of British strategy was confirmed by the events of the following two years, when against a background of almost continuous international tension east of Suez, one emergency after another dominated Whitehall's attention.

The overstretch, however, of British forces caused by the attempt to retain an independent global and limited-war capability, while at the same time not increasing the financial resources required for such a capability, led to the readjustment of strategic thinking between 1966 and 1968. Henceforth although Britain was to retain a small capability for limited extra-European operations, the main focus of attention was to be Europe, where it was recognized that both Britain's conventional and nuclear forces must be integrated more effectively into an alliance framework. The key word in British strategy was to be *interdependence* rather than *independence*. Increasingly the total-war theme was played down and the emphasis placed on deterring Soviet aggression in Europe by

*The concept of "broken-backed warfare" implies that in a nuclear conflict, the initial exchange might not be decisive, and war might continue at a lower level of intensity for some time afterward. The concept implies the necessity for making provisions for continuing operations after the initial exchange, particularly at sea. (See Darby, *British Defence Policy,* 47.)

coordinating nuclear and conventional components into an effective general and limited-war alliance strategy.

The priority given to Europe, however, and the decision not to retain a significant special capability for operations east of Suez were bound to cause problems if ever Britain's bluff were called. In 1982, as we have already seen, the Argentinian invasion of the Falklands islands could only be met by diverting substantial British forces from their priority roles in the Western alliance. The operation proved successful, but it imposed a certain material and financial strain on the British defense establishment. Had the Argentinians waited six months or even less, the weather and the effects of the 1981 review might have produced a different outcome. The debates following the war once again revealed the traditional dilemma in postwar British strategic thinking of how to reconcile preparations for a general war in an alliance context alongside a national capability for limited military operations.

## *Institutional Readjustment*

Parallel to, and partly responsible for, this constant process of adjustment and readjustment that has characterized British defense policy has been a continuing process of adaptation within the institutional structure for policy formulation. In 1946 the wartime expedient was put on a permanent footing when a minister of defense was appointed with responsiblity for coordinating the services. Beneath the minister came a series of committees, the most important of which were the Defence Committee of the Cabinet and the Chiefs of Staff Committee. This process of defense by committee was, however, sharply criticized in the postwar period for arriving at decisions based on compromise. Field Marshal Montgomery in his memoirs draws a picture of intense interservice rivalry for the scarce funds available. There was, he maintained, little overall coherent strategic planning, and no unified defense policy.[37] This was largely due to institutional weaknesses. In particular, the minister's powers remained inadequate to secure a unified policy, the the Defence Committee failed to perform its statutory function of integrating domestic, foreign, and military policies.

The first real attempt to modify the system in order to overcome the continuing problems of resource allocation came in 1955 when Sir Anthony Eden announced an expanded interpretation of the minister of defense's role, to include ensuring that the composition and balance of each service conformed to the government's policy.[38] Almost immediately the Suez debacle occurred, after which a new review of defense policy was initiated by Eden's successor, Harold Macmillan. Once again the powers of the minister of defense were extended. Duncan Sandys, the new incumbent, was authorized on 24 January 1957 to decide matters of general defense policy affecting the "size, shape, organization and disposal of the armed forces and their equipment and supply including defence research and development." These new powers of the minister of defense were confirmed in July 1958 in a white paper on the central organization for defense.[39]

This movement toward further centralization and utilization of his new powers by the minister caused resentment among the service chiefs, who saw their traditional autonomy and influence in the decision-making process gradually being eroded. As a result of their pressure the July 1958 white paper represented a compromise between the defense minister's inclination to carry the centralization of authority further and the service ministers' to maintain their independence. Although the overall authority of the minister was asserted, the Chiefs of Staff Committee retained its collective responsibility to the government for collective advice and its right of direct access to the prime minister. Sandys's attempt to confine the service ministers to the new Defence Board, which was to deal with purely military matters, was also overruled by the cabinet, and their attendance at the Defence Committee was left to the discretion of the committee.

The failings of this compromise solution became increasingly evident with the replacement of Duncan Sandys by Harold Watkinson in October 1959. The different personalities of the two men led to the realization of how little the changes that had been carried out under Sandys's direction had affected central control of defense policy formulation. The minister had been given explicit responsibility for the "formulation and general application of a unified policy," but he had not been given the necessary machinery to carry it out. The internecine struggles between the three services for appropriations as well as their power of initiative in policy formulation and in procurement remained as great as ever.[40]

Further attempts to readjust the machinery came in 1963 with the Thorneycroft-Mountbatten reforms.[41] On the prompting of the chief of defense staff, Lord Mountbatten, and Peter Thorneycroft, the incumbent minister of defense, the prime minister initiated an independent inquiry under General Ismay and Lieutenant-General Sir Ian Jacob into the organization of the defense machinery. The recommendations in their report (which was ready with astonishing speed in two months) became the basis of the white paper that was presented to Parliament in July 1963. A single Ministry of Defence was established under a secretary of state, absorbing the existing service ministries. In their place, signaling their reduced status, navy, army, and air force departments were established. Broad issues of defense policy would still be settled at cabinet level, but the Defence Committee would be replaced by a Committee on Defence and Overseas Policy. The Service Boards were to remain in being but only as subsidiary managerial instruments of a central Defence Council in which the ministers, the chiefs of staff, and the chief scientific adviser would deal with major policy questions. In general the reforms amounted to an administrative revolution in the field of defense policy formulation and a great leap forward toward greater centralization. The powers of the secretary of state were greatly enhanced, and the status of service ministers was drastically reduced, thus downgrading the independence and autonomy of the service ministries.

Further modifications and refinements occurred in 1967 when the service ministers were removed and replaced by two ministers, one for personnel and logistics and the other for research, procurement, and sales.[42] Again in 1970 these two ministerial posts were merged to form a single post for personnel, equipment, research and logistics.[43] At the same time, following the Rayner Report in 1971, a central procurement executive was established, and a minister of state for procurement was set up.[44]

In the early 1980s the single minister of state with three parliamentary secretaries (one for each of the three services) were replaced by two new ministers of state, one responsible for the armed forces and the other for defense procurement. The main aim of the new structure was to "increase the Secretary of State's ability to delegate functional responsibilities to his Ministers; to emphasize the defence, as opposed to the single service, responsibilities of the Department; to strengthen political direction throughout the Department and to allow far greater ministerial control of the defence procurement process at an earlier stage than in the past."[45]

At the same time the position of the chief of the defense staff (CDS) was strengthened. Although authority of the single-service chiefs of staff was maintained, the role of the CDS as the secretary of state's principal military adviser was further enhanced, in particular by strengthening his control over the central military staffs. As a result of the "consultative proposals" put forward by the defense minister, Michael Heseltine, in March 1984 further changes in the balance were envisaged. According to these proposals, the position of the CDS was to be strengthened to an even greater extent by providing him with a larger joint-service and civilian staff. At the same time it was proposed to reduce the staffs of the individual service chiefs and confine their responsibilties to the management of their services. Despite some support for this extension of centralized control over defense policy the proposals not surprisingly did cause some concern among the military establishment. Several ex-chiefs of the defense staff in particular expressed their reservations about the loss of service autonomy and influence over defense decision making. In a letter to *The London Times* Field Marshal Lord Carver argued that it was an error to imagine that policy and management were two different functions. They were, he suggested, "inextricably intertwined." He accepted that there has to be a central assessment and direction of the general balance of effort between different components of defense policy, but he went on to argue that

> . . . within that global central allocation of priorities and resources, the single Service machinery is best qualified to balance all the different factors affecting the effort to be provided by its service: to determine what weapons systems and organisation are required, how many units there should be, how they should be organised, equipped, trained and accommodated . . .[46]

According to Lord Carver, the permanent balancing act was highly sensitive to a large number of factors, which could only be accurately assessed as a result of "accumulated experience and contact with the grass roots." Such experience was only to be found within the single services. As far as Lord Carver and several of his colleagues were concerned, the Heseltine proposals were likely to undermine the ability of the service chiefs to advise the CDS and relate them to the "task of routine management of their services."[47]

Since 1945, therefore, there has been a continual process of adjustment within the central organization of defense. This has meant that the Ministry of Defence and the higher echelons of the services have been in continuous flux when they have had to carry out reappraisals and make constant adjustments to British defense policy. Plans have had to be revised to conform to new decisions when the defense structure and the chain of command was constantly melting and reforming. This constant process of adaptation has caused British defense policy in the postwar period to be not always distinguished by clarity or consistency of purpose.[48]

This conclusion, however, in no way detracts from the value of the defense establishment that was maintained during the period. The successful completion of numerous military operations testifies to this fact. Notwithstanding the problems caused by interservice rivalries and the failure of the overseas base policy, the military achievement was considerable. Despite the lack of clear political direction and the limited resources available, the services adapted to the changing environment with great alacrity. The preoccupation in Whitehall with the nuclear deterrent did not prevent the services from developing new approaches to the problems of overseas security. The operational record of the services attests to their success in developing an effective system for the provision of military support for British foreign policy in the post-colonial era.

## *Current and Future Issues*

In some important respects the process of adjustment for the services and for defense policy in general, which has characterized much of the postwar period, is likely to continue. While changes in the domestic and international environments that might alter significantly the nature of British defense policy cannot be foreseen, it seems likely that for the rest of the 1980s and longer, Britain's security interests will be concentrated in Europe. Although the main direction of policy may have been decided upon, many other significant choices will nevertheless have to be made within the confines of the general parameters of the European defense orientation.[49]

One of the major questions for the future is whether Britain's security interests will be concentrated in a solely European defense organization, or whether they will continue within an Atlantic alliance in more or less its present form. In recent years, as Britain has built up closer political and economic ties with other European countries, it has also, for obvious political reasons, been moving increasingly toward greater identification with the European caucus within the alliance. Apart from Britain's membership of the Eurogroup and the Independent European Program Group (IEPG) various studies produced by officials within the EEC in the 1970s (such as the Tindemans Report) advocated progress toward the kind of harmonization in the defense field that was taking place in various foreign policy fields. At the same time economic pressures have also been pushing Britain toward greater defense integration with her European allies. The major determinant of British defense policy in the postwar period has been, and still is, one of finance.[50] With limitations placed on the percentage of GNP spent on defense and the escalation of weapons costs, it was perhaps only natural that Britain should seek closer collaboration in weapons procurement with other European states. The most important of these projects has been the Jaguar close-support-and-attack aircraft produced by Anglo-French companies, and the Tornado, produced by three nations—Britain, Italy, and West Germany. This tentative move toward joint procurement has led many to push for the establishment of a more effective European procurement agency to integrate weapons development for the European members of NATO as a whole.[51] Some commentators have gone even further and argued that because of the need to harmonize European defense and foreign policies at a time of increasing uncertainty over the scale of the long-term American presence in Europe, Western European states should form a European defense community to coordinate European defense efforts and supplement their own security.[52]

Such a project, however, poses a number of fundamental difficulties. Firstly, and perhaps most important, there is the problem of establishing a European defense community that would be able to give the kind of political direction to coordinate the national defense policies of Western European states before some form of political union is in fact achieved. The question is whether such a European defense organization would be a spur to further political cohesion or whether, as perhaps is more likely, political union is a prerequisite for an effective European defense community in the first place.[53] On top of this obvious political problem there is also the difficulty of the form and composition of such an organization. Should it be based on the old Western European Union, outside

NATO and including France, or should it build on the foundations of the Eurogroup within NATO and exclude France? Both organizations or any other European defense arrangement that could be envisaged would face the same problem of its relations both with NATO and with the EEC.

For Britain the dilemma this creates is particularly acute. Largely for political reasons there is an incentive to forge stronger ties in defense with other EEC members, particularly the French. This feeling has been enhanced by uncertainty over the direction of American foreign policy in the late 1970s and early 1980s (including the ambiguous support given to Britain in the Falklands crisis and the lack of consultation in the Grenadian intervention). On the other hand, traditional friendship and ties with the United States as well as the recognition of the necessity of the American nuclear guarantee for Europe would seem to suggest a continuing emphasis on the Atlantic alliance. The problem for the government therefore is how far to move toward increased European defense integration, desirable for economic and political reasons but possibly detrimental to the defense relationship with the United States.[54]

In one area of that defense relationship at least a close and preferential partnership has been reconfirmed in recent years. After a long period of consideration over the replacement for the Polaris nuclear force as we have seen the government announced its decision on 15 July 1980 to purchase Trident missiles from the United States. Despite widespread speculation that Britain could not expect access to the latest American strategic delivery systems, especially for a cheap price, the Trident Accords of 1980 and 1982 were agreed to on the same favorable financial terms as the Polaris Sales Agreement of 1963.[55] In the nuclear field at least the continuation of the "special relationship" was regarded as crucial to British security interests. Indeed, the Trident Agreement, if implemented, would tie Britain closely to American strategic interests well beyond the year 2000.

For some observers of the defense scene in Britain the Trident deal highlights a major paradox in contemporary defense policy and poses serious problems for the future. The paradox arises from the fact that one of the major reasons put forward in support of the maintenance of a strategic nuclear force was the uncertainty about American protection. The arrival of strategic parity between the superpowers and the major difficulties of European-American relations in the early 1980s, together with the specific events surrounding the Falklands war, reinforced the belief in certain quarters in Britain that a last-resort weapon is of crucial importance to British strategic interests in the dangerous years ahead. To be sure, the strategic relationship helps to bind the two countries together, but the possession of the most sophisticated weapons system yet developed is also designed to help Britain survive in a world in which the United States has gone its own way.

There is also the problem, however, that although Trident missiles will be acquired on very favorable financial terms, they remain expensive for a country with Britain's financial problems. As such there are grave doubts about whether Britain can afford Trident and provide the kinds of conventional forces necessary to contribute to NATO defenses and the kind of intervention forces necessary to defend British and Western interests outside the alliance.[56]

Apart from the extra budgetary problems the Trident accords have helped to generate, the nuclear question itself has become an important issue in British politics in recent years and is likely to remain so in the next few years. After the virtual disappearance of the Campaign for Nuclear Disarmament (CND) in the 1970s it has reemerged, phoenix-like, to become a significant organization of protest in the 1980s. The deterioration of East-West relations in the late 1970s and early 1980s, which coincided with the announcement of the deployment of new weapons systems by Western governments, helped rekindle much the same kind of fear of the possibility of nuclear war that gripped the populations of many Western European states in the late 1950s. In the early 1980s the decision to replace Polaris by the more sophisticated Trident system and the decision to deploy 160 American cruise missiles at Greenham Common and Molesworth led to a lively debate inside and outside Parliament. By 1982 all the opposition parties were against the purchase of Trident, and the Labour party had committed itself to getting rid of all nuclear weapons and the closing down of American nuclear bases in Britain. This breaking down of the traditional bipartisan approach to defense in Britain and growing public involvement in the debate about alternative defense strategies suggest that defense issues are likely to remain politically important in the years ahead. Despite the Conservative party victory in the 1983 election and the unpopularity of Labour defense policy with the electorate, the Labour party has continued to criticize the nuclear bias of Prime Minister Thatcher's defense policy and to advocate its own non-nuclear defense policy.[57] The ongoing defense debate and especially the outcome of the 1987–88 election is almost certain to have a profound impact on the orientation of the British defense posture in the late 1980s and beyond.* The next election in particular could well decide whether Britain retains a close defense partnership with the United States; continues to cooperate in security matters with her European allies; maintains a strategic and tactical nuclear capability; and continues to provide balanced conventional forces.

Even before that election, however, as we have seen, budgetary problems are likely to push the Thatcher government toward taking crucial decisions about priorities by the mid-1980s. In 1981, the secretary of state for defense, John Nott, argued in his white paper, *The Way Forward*, that Britain could not "go on as before"; something had to be cut.[58] Subse-

---

*This assumes that the Thatcher government remains in power for all or most of its term of office.

quently the Falklands campaign led to a partial revision of that policy and a decision to continue, for a time, the present scale of surface naval forces and to improve Britain's intervention capabilities.[59] At the same time the government decided not to continue the 3 percent increase in real terms per annum beyond 1986.[60] As a result the government was attempting to do even more than in 1981 with less resources.[61] The inescapable conclusion seemed to be that yet another review would have to take place in the later 1980s with yet another round of cuts.[62] Thus, the process of adjustment and readjustment is likely to continue to be the order of the day.

## NOTES

1. See Phillip Darby, *British Defence Policy East of Suez 1947–68* (London: Oxford University Press, 1973), 16–21, 135–42.

2. This was particularly the case over Singapore in the years before the Lancaster House Conference in April 1956 and also over Kenya a few years later. See Darby, ibid., 205.

3. See C. H. Bartlett, *The Long Retreat: A Short History of British Defence Policy 1945–1970* (London: Macmillan, 1972), xi.

4. See Harold and Margaret Sprout, "Retreat from World Power: Processes and Consequences of Readjustment," *World Politics* 15, no. 4 (July 1963).

5. See Andrew J. Pierre, *Nuclear Politics: The British Experiment with an Independent Strategic Force 1939–1970* (London: Oxford University Press, 1972), 71. See also H. C. Deb., Col. 239 (5 March 1946).

6. Pierre, ibid., 71.

7. One of the main proponents of this idea was Lord Alanbrooke. See Bartlett, *The Long Retreat*, 11; *The Economist*, 8 February 1947 and 27 March 1948; *Royal United Service Institute Journal* (RUSIJ) (no. 566, May 1947): 182–86.

8. See R. Osgood, *NATO: The Entangling Alliance* (London: University of Chicago Press, 1962).

9. The fifty-year Brussels Pact of mutual collective assistance against any aggressor was signed on 17 March 1948 by Britain, France, the Netherlands, and Luxembourg. See J. Baylis, "Britain and the Dunkirk Treaty: The Origins of NATO," *Journal of Strategic Studies* (June 1982), and "Britain, the Brussels Pact and the Continental Commitment," *International Affairs* (Winter 1984).

10. For the interwar debate see M. Howard, *The Continental Commitment* (London: Temple Smith, 1972), 96–120; see also Bartlett, *The Long Retreat*, 20–21; the Lisbon goals involved an attempt to build up a force of ninety-six divisions in Europe by 1954 (see Osgood, *NATO: The Entangling Alliance*, 87; and see S. Robinson, "The Global Strategy Paper of 1952," unpublished M.Sc. thesis, University of Wales, 1974).

11. Cmnd. 3540, *Statement on the Defence Estimates*, 1968; Cmnd. 3701, *Supplementary Statement of Defence Policy*, 1968.

12. C. Barnett, *Britain and Her Army 1509–1970* (London: Pelican, 1974), 479.

13. See R. L. Clutterbuck, *The Long, Long War* (New York: Praeger, 1966).

14. See G. Blaxland, *The Regiments Depart* (London: Kimber, 1971), 375–410.

15. See D. Rees, *Korea: The Limited War* (London: Macmillan, 1964).

16. J. Howarth, "The Kuwait Operation," unpublished M.Sc. thesis, University of Wales, 1972.

17. See A. J. Barker, *Suez: the Seven Day War* (London: Faber and Faber, 1964).

18. Cmnd. 8758, *The Falklands Campaign: The Lessons*.

19. For a questioning of the 1981 review in the light of the Falklands campaign, see J. Baylis, ed., *Alternative Approaches to British Defence Policy* (London: Macmillan, 1983).

20. For the extensive responsibilities of Indian troops for empire defense see Darby, *British Defence Policy,* 2; Sir Anthony Eden, *Full Circle* (London: Cassell, 1960), 370–75; and for Sandys's reappraisal see Cmnd. 124, *Defence: Outline of Future Policy,* 1957 (see also A. Pierre, *Nuclear Politics,* 96).

21. M. Howard, *Central Organization for Defence* (London: RUSI, 1970), 9.

22. See Patrick Gordon Walker, *The Cabinet* (London: Fontana, 1970), 124–34.

23. Charles Douglas-Home, "Concentration of Strength," *London Times,* 23 January 1968.

24. Cmnd. 5976, *Statement on the Defence Estimates 1975.*

25. See John Baylis, ed., *British Defence Policy in a Changing World* (London: Croom Helm, 1977), 27.

26. Cmnd. 8288, *The United Kingdom Defence Programme: The Way Forward.*

27. See M. Chichester and J. Wilkinson, *The Uncertain Ally: British Defence Policy 1968–1990,* (London: Gower, 1982), and J. Baylis, ed., *Alternative Approaches to British Defence Policy;* and Cmnd. 8758. The 1984 white paper, Cmnd. 9227, slightly improved the position for the navy. Up to eight frigates and destroyers, which were to be mothballed in the stand-by squadron, are now to remain in service with the fleet.

28. See Darby, *British Defence Policy,* 168–72.

29. Cmnd. 124, *Defence: Outline of Future Policy.*

30. Darby, *British Defence Policy,* 68.

31. See Pierre, *Nuclear Politics,* 101–11.

32. Ibid., 275–82.

33. For a discussion of the Polaris Improvement Program and Chevaline see L. Freedman, *Britain and Nuclear Weapons* (London: Macmillan, 1980).

34. David Greenwood, "The Trident Programme," *Aberdeen Studies in Defence Economics,* no. 22 (1982).

35. Cmnd. 9075, *Statement on Defence,* 1954.

36. See Cdr. W. J. Crowe, "The Policy Roots of the Modern British Navy, 1946–63," unpublished Ph.D. dissertation, Princeton University, 1965.

37. Field Marshal Montgomery, *Memoirs* (London: Collins, 1958), 487–97.

38. Eden, *Full Circle,* 375.

39. Cmnd. 476, *Central Organization of Defence,* 1958.

40. See L. W. Martin, "The Market for Strategic Ideas in Britain: The Sandys Era," *American Political Science Review* 56, no. 1 (March 1962): 23–41.

41. Cmnd. 2097, *Central Organization for Defence,* 1963.

42. Cmnd. 3203, *Statement on the Defence Estimates,* 1967.

43. Cmnd. 4290, *Statement of the Defence Estimates,* 1970.

44. Cmnd. 4641, *Government Organization for Defence Procurement and Civil Aerospace.*

45. Cmnd. 8529/1, *Statement on the Defence Estimates 1982.*

46. "Letters to the Editor," *London Times,* 23 March 1984.

47. See also the letters to the editor from Admiral of Fleet Lord Lewin, Admiral of the Fleet Lord Hill-Norton and Admiral of the Fleet Sir Henry Leach in *The London Times,* 20 March, 30 March, and 7 May 1984.

48. See L. W. Martin, "The Long Recessional," *Adelphi Paper,* no. 61 (London: International Institute for Strategic Studies, 1969).

49. For an analysis of the implications of membership of the EEC on British defense policy see J. Baylis, "The Defence Implications of British Membership of the EEC," *The Royal Air Forces Quarterly* (Spring, 1974); and J. C. Garnett, "European Security and an Enlarged Community," in *The Defence of Western Europe* (London: Macmillan, 1974).

50. See David Greenwood, *Budgeting for Defence* (London: RUSI, 1972).

51. See S. W. B. Menaul, "Britain in Europe: The Defence Aspect," *Defence,* 2, no. 9 (Sept. 1971): 29.

52. See W. Schutze, *European Defence Cooperation and NATO*, (Paris: Atlantic Institute, 1970).

53. For a discussion of this see Garnett, "European Security."

54. It is of interest that in 1984 Britain initially tried to strengthen the IEPG but then gave its cautious backing to a new Franco-German initiative to revive the WEU. See *The London Times*, 13 June 1984.

55. See J. Baylis, *Anglo-American Defence Relations 1939–1984: The Special Relationship* (London: Macmillan, 1984).

56. See David Greenwood, "The Trident Programme."

57. In January and February 1984 the new Labour leader, Neil Kinnock, confirmed that this would remain Labour party policy. See *London Times*, 23 January and 6 February 1984.

58. Cmnd. 8288. On this occasion a cut in the size of the surface fleet.

59. Cmnd. 8759. Much will depend on how long Britain has to maintain a sizable naval force in the South Atlantic. Some further revisions in Britain's surface naval forces were announced in Cmnd. 9227 in 1984.

60. The government's *Public Expenditure Survey Report*, published in February 1984, envisaged little more than 1 percent real increase in 1986–87 and even less in the years ahead.

61. The problem was compounded by the significant rise in the price of Trident as a result of the changes in the exchange rates in 1983–84.

62. See John Baylis, ed., *Alternative Approaches to British Defence Policy*. For the debate over whether another review is inevitable, see *The London Times*, 28 February 1984.

# 6

# *French Defense Policy*

JOHN BAYLIS

One of the most crucial questions in French defense policy in the period since the Second World War has been the issue of continuity and change. Speaking at a press conference in October 1966, President de Gaulle highlighted the contrast that had existed in French foreign and defense fields for much of the postwar period between the changes and detours of policies and the essential consistency of the central objectives governing them.[1] Although French policies in the Fourth (1946–58) and Fifth (1958– ) republics had been constantly adjusted to meet the exigencies of domestic and international circumstances, de Gaulle argued that the major goals of policymakers had remained the constant search for the independence and sovereignty of France and the re-establishment of the "grandeur" of the nation. Policies may have been in a continual state of flux, but French motivation had remained unchanged.

Although the president was undoubtedly correct in this view, he failed to tell his audience that the consistent search for greatness and independence had been (and still remains) the source of a fundamental dilemma for French policymakers for much of the period since 1945. French statesmen have framed their policies in the light of the aspiration, and indeed the illusion of "grandeur,"* only to be confronted with the stark reality of the reduced importance and shrinking resources of their nation. It is perhaps not surprising that the achievements of French foreign and defense policy, although impressive, have not always fulfilled the exalted visions of statesmen like de Gaulle. The attempt to balance resources and aspirations remains as difficult for French defense planners in the 1980s as for their predecessors throughout the period since the Second World War.

---

*The policy of grandeur was thought of partly in terms of independence but also in terms of the provision of the material foundations not necessarily to match the other superpower but certainly to rival them.

## *Foreign and Defense Policies*

In France for much of the postwar period the relationship between foreign and defense policies has been a particularly close one.[2] For leaders of the Fourth Republic and, more particularly, the Fifth Republic, defense policy has been seen as one of the most important instruments for achieving the objectives of French foreign policy. Military forces have been envisaged by successive French governments not only as a means of reestablishing France's prestige and status in Europe and the world, but also as a symbol of the nation's independence of both allies and enemies alike.

France's best example of this is its development of nuclear weapons. Even before France's first atomic explosion in 1960, supporters of the military nuclear program in both the Fourth and the Fifth republics saw nuclear weapons as an important instrument to enhance the nation's political status and military capability. There was, however, a major difference in interpretation between the two republics on the role that nuclear weapons were designed to perform. For the leaders of the Fourth Republic, atomic weapons were seen as providing support for French policy within the existing framework of the North Atlantic Treaty Organization. It was envisaged that nuclear weapons could provide a base for a more independent national policy and that they could enhance France's standing and prestige, but in both cases this was to be within the Atlantic alliance. For de Gaulle, however, the development of atomic armaments was designed primarily to support "his independent foreign policies, which sought to change the European and international system and France's role in it." The *force de frappe,* France's nuclear strike force, was seen as a tool that could be used to pursue the president's broader foreign policy objectives, which were much "bolder and more grandly conceived than those of French leaders before 1958."[3] These bolder objectives are summarized by Wilfred L. Kohl as being:

> the restoration of French grandeur through the achievement of a global role for France and—so far as possible—coequal status with the United States and Great Britain; the subordination of West Germany to France and the preservation of French leadership in Western Europe; reunification of the two halves of the European continent in a loose association of states "from the Atlantic to the Urals," and an independent role for this "European" Europe in world politics.[4]

To General de Gaulle, nuclear weapons were more important from a diplomatic than a strategic angle. Such weapons were seen as a *"jeton de présence"* (token of presence) among the superpowers of the world. They were a *force de persuasion* that would enable the voice of France to be heard and that would secure it a seat at the bargaining table with the other great powers.[5] The identification of France with other members of the select nuclear club would conceal the enormous disparity between them in terms of military strength while at the same time emphasizing the qualitative superiority of France over other non-nuclear states, which justified her presence in the negotiations in which the major problems of mankind were discussed.

The policy of reestablishing France to her "rightful" position in international affairs also involved the search for equality of status within the Atlantic alliance with the other two nuclear powers, the United States and Great Britain. To gain such equality, de Gaulle, like the later leaders of the Fourth Republic, realized that France required a nuclear capability similar to that of the Anglo-Saxon states. To achieve this objective the French president (again like his predecessors) was not adverse to entering into collaboration with the United States.[6] Consequently with the amendment to the U.S. Atomic Energy Act in 1958, which gave preferential treatment to Great Britain, the authorities in Paris sought similar treatment for France.

De Gaulle was to find, however, that nuclear sharing was something that the American government was only prepared to countenance with its closest of allies, Great Britain. On a number of occasions when the prospects of cooperation seemed good, American congressional opposition prevented agreement. In July 1958 after John Foster Dulles's visit to Paris, a promise was given of American help in the building of a submarine reactor. The offer, however, had to be withdrawn after objections from the Joint Congressional Committee on Atomic Energy. Again in 1959 negotiations between French and American firms to produce a French intermediate-range ballistic missile (IRBM) were cut short by an agreement between the State and Defense departments to prohibit all cooperation between the firms involved. Similarly in 1962 when Secretary of Defense Robert McNamara and General Maxwell Taylor, chairman of the Joint Chiefs of Staff, recommended a sharing of atomic information with France after discussions with Pierre Messmer, the French minister of the armed forces, the recommendation was not taken up by the American government.

The disappointment and resentment caused by these American responses served to reinforce de Gaulle's view that France should push ahead with her own nuclear program. The failure to get American assistance, particularly in light of the preferential treatment given to Britain, served likewise to confirm the French view of Anglo-Saxon predominance within the alliance and in so doing strengthened the determination and conviction of Charles de Gaulle to create a national nuclear force that would have as one of its primary purposes the freeing of France from strategic dependence on the United States.*

---

*The United States had also entered into agreements with Canada, West Germany, The Netherlands, Turkey, and Greece providing them with some nuclear information regarding the training of personnel and the stockpiling of nuclear explosives in these countries. A similar treaty was not signed with the French until 1961.

The French nuclear program also held an important place in French relations with Germany and the determination of both Fourth and Fifth republican governments to avoid subordination to the new West German state. The question of the military uses of nuclear power was first raised in the French cabinet on 26 December 1954 at a time when questions of German rearmament loomed large in political circles in France. Mendès-France, who brought the issue to the cabinet and who at the same time made the decision to launch a secret program of studies and preparations for a prototype of a nuclear weapon, did so because of his fear of domination by a rearmed Germany within the Western Alliance.

From this period on "the theme of keeping ahead of Germany runs like a thread through subsequent debates on foreign and defence policies" in France. This is well illustrated by the discussions that took place over the establishment of the European Atomic Energy Community (EURATOM). During the negotiations the French insisted on modifying a clause in a draft of the treaty that would have required her to renounce the fabrication of atomic weapons. France could thereby develop a military nuclear capability while the same right was denied to the German government.[7] The leaders of the Fourth Republic, therefore, and later the Fifth Republic, have shown themselves willing to accept partnership with Germany provided that France was the senior member. The best way to achieve this superiority was seen to be through the development of a national nuclear force, the creation of which was denied to Germany.

French partnership with Germany, which was symbolized by the French-German treaty of January 1963 together with the possession of a national nuclear force, was also the basis of de Gaulle's long-range pan-European design. From 1960, de Gaulle increasingly began to develop his ideal of a "European *entente* from the Atlantic to the Urals" which would enable Europe to solve "its problems peacefully, particularly that of Germany, including reunification, and to attain, inasmuch as it is the main hearth of civilization, a material and human development worthy of its resources and its capacities."[8] To de Gaulle, France was the only European power capable of taking the initiative to achieve the grand design of reuniting the two halves of Europe to enable the Continent to play a major role in world affairs independent of the two superpowers, America and Russia. In the short term before an all-European settlement could be achieved, however, France would be responsible for taking the coordination of West European states a stage further.

In 1959 therefore, the French president put forward a new initiative in the form of the Fouchet Plan designed to achieve increased West European political cooperation that would have "started Europe along the path toward European confederation." From de Gaulle's point of view, support for a political union of Western European states represented an attempt not only to unify Europe but also to coordinate European and more particularly German, foreign, and defense policies under the leadership of France and to "achieve a consensus on a more independent

European stance in security matters."[9] Again the focal point of this objective was the French nuclear force. As part of his plan, the general argued for the establishment of a European defense arrangement, which would be built around the nucleus of the French atomic force. The rather vague implication of this proposal was that French nuclear forces would be placed at the disposal of Western Europe when West European political integration reached a sufficient level. The lack of an indication of French willingness to share control of her nuclear force, however, led to the breakdown of these Fouchet discussions in 1962. France's partners soon realized that de Gaulle's proposals were directed more at enhancing the French position in Western Europe than at developing a new role for Europe in a revised international system.

Despite the failure of the Fouchet Plan, the national nuclear force was again at the heart of a new French initiative in the mid-1960s designed once again to break away from the status quo in Europe represented by the two alliance systems dominated by the Soviet Union and the United States. On this occasion the *force de frappe* served to underpin French withdrawal from the NATO military command in 1966. Once again this move was aimed at replacing the Western defense system based on the Atlantic alliance by a new security framework that would embrace the European continent as a whole. This time the withdrawal from the military structure of NATO was synchronized with a policy of rapprochement with the Soviet Union and Eastern European states. The Soviet government, however, was unreceptive to the new French initiative, and in 1968 with the Russian invasion of Czechoslovakia and growing domestic problems, de Gaulle began, to a certain extent, to modify the independent line implicit in the policies adopted in 1966.

In both the Fouchet Plan and the design in the midsixties for a pan-European security arrangement, the French possession of a nuclear capability was of critical importance. The Force Nucléaire Stratégique (FNS) was perceived as an instrument designed principally to serve the Gaullist policy of reestablishing France as a great power in Western Europe and on the international stage as well as securing the independence of the nation from domination by the United States. The pursuit of such a policy, however, involved a fundamental dilemma for French leaders. With the limited resources available, France was not able to play the world role envisaged by those like de Gaulle. Only an independent united Europe could rival the other two superpowers on the same terms. To gain the power base needed to play the role it aspired to, therefore, France would have had to pool its resources with the other European allies and allow itself to become integrated into a European political union. In so doing France could then perhaps have been part of a great independent bloc. But in the process the nation would also have lost its independence and sovereignty so highly valued by French leaders, particularly in the military field. Given France's limited national power base, therefore it was unlikely that it could be both "great" and "independent." Nevertheless the attempt to have it both ways has been very much a characteristic of French foreign and defense policies.

## *National Control and Defense Integration*

The Second World War provided contradictory lessons as far as French policymakers in the postwar period were concerned. French liberation by the Allies emphasized the efficacy of defense alliances and the need for France to cooperate closely in the military fields with other states if renewed threats to its security materialized. But the experience of the first part of the century, as well as other events in the Second World War, suggested an alternative interpretation. As Wolf Mendl points out: "Neither the alliance system before World War I nor the attempt to construct a system of collective security between the wars, saved France from great suffering and a feeling that she had been used as an advance pawn and as a source of manpower in terms of the peripheral strategy of the Anglo-American powers."[10]

Such a perception of past experiences led naturally to the feeling that in the last resort France would have to provide for its own security. French interests were unlikely to coincide exactly with those of its allies, and in the vital sphere of defense it must retain as great a degree of self-sufficiency and independence as possible.

The contradiction between these two attitudes has remained at the heart of French policies toward alliances since 1945. Initially in 1946 the new premier, Charles de Gaulle, attempted to strengthen French security vis-à-vis Germany and at the same time to reestablish French independence of Britain and the United States by signing a treaty with the Soviet Union. Such a demonstration of independence, however, was short-lived. Increasingly, as relations between the West and Russia deteriorated, French political, economic, and military weakness forced it to depend more on the Anglo-Saxon states. In 1947 the Dunkirk Treaty was signed with Great Britain against the possibility of future German aggression. Similarly in 1948, France joined the Brussels Pact, which, although directed mainly at the Soviet Union, was also largely on French initiative, designed once again to forestall potential German aggressive moves.[11]

With this increased emphasis on cooperation with allies, however (taken an important stage further with the signing of the NATO Treaty in April 1949), the defense planners in France had to determine the degree of defense integration with allies that would give the needed security but that would not be detrimental to the continuation of national sovereignty and control over defense policy. The problem was made more complicated when the deterioration of the international situation caused by the

Berlin blockade and the Korean War led the American government to pressure its allies to accept West German rearmament to reinforce the Western defense effort. Continuing suspicions of Germany caused the French government to put forward the so-called Pleven Plan, which called for the creation of a European Defense Community (EDC) in which German forces would be allowed to form part of a European army but would not be under the German government. For the integrationists in France in particular, the EDC would serve the cause of European unity and also would enhance French security against Germany and Russia. From the beginning, however, the plan was severely criticized, especially by those who saw it as a threat to national sovereignty. On 30 August 1954 the treaty, which had been put forward by the French government, was rejected by the French National Assembly. Instead a compromise was accepted on the basis of the British-sponsored Paris Agreements whereby the Western European Union (WEU) was extended to include Germany and Italy. Britain also agreed to station troops on the Continent until 1994, in part at least to alleviate French apprehensions of German rearmament.

Although toward the end of the Fourth Republic increasing criticisms were made in France of dependence on the United States in defense, particularly after the American refusal to help France in Indochina prior to the defeat of Dien Bien Phu and the active hostility at the time of the Suez crisis, Fourth Republican governments nevertheless sought to enhance France's political status and military role within the existing framework of the North Atlantic alliance. The creation of the Fifth Republic under de Gaulle in 1958, however, led to a new twist in French attitudes toward the question of integration or self-sufficiency in defense. The fundamental importance of French sovereignty and the need to enhance France's status caused the new president to question the structure of NATO and American domination of the organization.

De Gaulle's criticisms came to rest on two important arguments. On the one hand he believed that the launching of the Soviet Sputnik in October 1957 had greatly downgraded the credibility of the American nuclear guarantee to Europe. The Soviet capability of attacking the American homeland suggested to de Gaulle that it was increasingly unlikely that an American president would risk the devastation of the United States to save Europe. On the other hand, he also saw the risk that events in the Middle East and the Formosa Straits crisis in the late 1950s had demonstrated, that France might be dragged into a war by the United States. A conflict between the United States and the Soviet Union over which the French government had no control would almost certainly, de Gaulle argued, escalate into a war in which France was involved regardless of the fact that its vital interests might not be at stake.

The French president's dissatisfaction with the existing structure of NATO led him to send his famous secret memorandum to President

Eisenhower and Prime Minister Macmillan in September 1958. In his note to the two leaders, de Gaulle suggested the creation of a new tripartite organization consisting of Britain, France, and the United States that would take "joint decisions" on all global questions affecting their security interests. "France," de Gaulle argued, "was ready to resume its historic role in world affairs" alongside the other great powers.[12] He also threatened that France's subsequent cooperation in NATO would depend on the acceptance of his demands for the tripartite organization.

A certain amount of evidence suggests that de Gaulle's directorate proposal was a tactical ploy on his part, which, once rejected by Great Britain and the United States, could then be used to justify his subsequent actions in progressively reducing France's participation in NATO.[13] Whether or not the general had a detailed blueprint for action is difficult to establish. He does, however, point out in his memoirs that he did not expect a favorable response to his proposal and that the memorandum was the key to his plan for a step-by-step disengagement from NATO, "an objective already clearly formulated in 1958."[14]

Subsequent French policies toward NATO would also seem to reinforce this view. In December 1958, following what he considered to be the unsatisfactory response by the United States to his proposals, de Gaulle announced his unwillingness to integrate French tactical aircraft into a European air defense command.[15] This was followed in March 1959 by the withdrawal of French naval units that had previously been committed to NATO's Mediterranean command. The most far-reaching attack so far against the organization, however, was delivered by the president in a speech at the Ecole Militaire on 3 November 1959 when he refused to accept the NATO tenet of "integration" and emphasized his sacred principle of national defense. "It is necessary," he argued, "that the defence of France be French . . . the concept of war or even of a battle in which France would no longer be herself and would not act on her own accord, following her own goals, such a concept cannot be admitted." France could combine its defense with that of other countries and associate its strategy with others, but integration could not be accepted: "each country must play its own part."[16]

The ideas expressed in this speech in 1959 have remained at the core of French defense policy ever since. In 1966, with the perception of a much-reduced Soviet threat, de Gaulle considered it opportune to take his policy to its logical conclusion and withdraw from the military structure of NATO. The subsequent movement toward neutralism designed to emphasize French independence in foreign policy, however, had to be curtailed to a large extent due to the combination of domestic and international events of 1968–69. The Soviet invasion of Czechoslovakia in 1968 and the domestic and monetary crisis of 1969 caused de Gaulle to temper his aspirations and in part to reappraise his attitudes toward

French relations with the other NATO members. Just before the general left office there were signs that France was once again interested in closer joint nuclear and military cooperation within the alliance. Renewed French interest in further cooperation was shown in a lecture in March 1969 given by Chief of Staff General Michel Fourquet at the Institute of Higher Defense Studies.[17] In his speech General Fourquet argued that the task of containing an enemy attack short of resorting to tactical or strategic nuclear weapons could be accomplished more effectively and efficiently through alliance action rather than through discrete attempts by the several states themselves. The implication of what Fourquet was saying seemed to be that given its limited resources France could only defend itself particularly at the conventional level through a much greater degree of cooperation with its allies than had been characteristic of French defense policy since 1966.

It is important to note, however, that although General Fourquet was arguing for greater cooperation with NATO than in the past, he was *not* suggesting a return to military integration within the alliance. Indeed, despite the recognition in the subsequent 1972 defense white paper that "France lives in a network of interests which go beyond her borders,"[18] French concern about the reliability of the American guarantee to Europe, its rejection of American domination of the alliance, and an emphasis on independence continued to sway defense thinking during the early seventies. The French strategic concept remained largely one of *sanctuarisation totale* (total sanctuary or "fortress France"), which had been developed in the late 1960s.

For many observers in France, however, these traditional Gaullist views about alliances were seriously threatened by the 1976 *loi de programmation* of President Giscard d'Estaing. In the preamble to the new six-year plan (covering 1977–82) the president argued: "It would indeed be illusory to hope that France could maintain more than a reduced sovereignty if her neighbors had been occupied by a hostile power or were simply under its control. The security of Western Europe as a whole is therefore essential for France."[19] This, together with a policy of increasing (if selective) cooperation with NATO brought substantial criticism from Gaullists and Communists alike. To both right and left, France seemed to be moving step by step back toward rejoining NATO.

Such a view was further confirmed according to government critics by changes in strategy announced by the Chief of Staff General Méry in a speech to the Institute des Hautes Etudes de la Défense Nationale in March 1976.[20] In this speech General Méry rejected the strategy based on *sanctuarisation totale* and instead advocated a strategy of *sanctuarisation élargie* (extended sanctuary). The latter, Méry argued, would allow France to intervene with the whole, or part, of its forces "throughout the entire zone where the security of its territory might be most threatened."

Despite the political furor that greeted the 1976 defense plans, particularly from Gaullists, it is very debatable whether they represented a wholly radical change in French defense thinking. In the debate that followed the publication of the *loi de programmation,* government officials, largely for domestic political reasons, went out of their way to stress the continuity with the past. General Méry in particular pointed out that although greater collaboration with NATO allies was desirable, it was "completely out of the question" that France "should undertake in advance—in peacetime—to occupy a space in the framework of the Allied strategy for the defense of the forward area." President Giscard d'Estaing likewise emphasized that there "can be no question" of France rejoining NATO.[21] To both men national control in decision making remained at the core of French defense policy. They were at pains to point out, however, that national control did not preclude cooperation. To Méry it was "a question of independence of decision, not necessarily leading to autonomy in action." To Giscard there was "no contradiction between belonging to an alliance and pursuing an independent policy." After all de Gaulle himself had distinguished between "integration" and "cooperation." It also seems to be true, as Jean Klein has pointed out, that "the 1976 statements by Giscard and Méry were not very different from those made by Armed Forces Chief of Staff Michel Fourquet in 1969."[22]

There are other aspects of the 1977–82 *loi de programmation* that must be considered before an overall judgment can be made about its significance for French defense policy. On the question of alliances, for example, the official statements suggest a certain ambiguity—indeed, some have said a deliberate ambiguity.[23] Some pronouncements have emphasized narrow Gaullist views while others have suggested a distinctive change in tone and direction. The conclusion seems to be that basic French attitudes toward alliances have not changed. Such a judgment, however, needs some qualification. There does at least seem to be some modification in attitude. This seems to be reflected, as David S. Yost has argued, in "a greater concern about the security of France's West European allies, and a somewhat greater willingness to prepare for possible action in conjunction with NATO, in the event of war."[24] In this respect French policy in the later 1970s was characterized by increased pragmatism. This is the policy that has also been extended further by the Socialist government of President Mitterand in the early 1980s. Gone are the more extreme notions of neutralism and total independence in defense. Important sediments of these ideas remain, but with the range of collaboration that has increasingly been established with NATO countries, especially West Germany, French policies do seem to have been progressively adjusted and refined to take rather more account of the perceived realities of the threats facing France.

## *The Relationship between Conventional and Nuclear Forces*

A problem closely related to the debate between defense integration or national control and one equally affected by the central dilemma facing policymakers has been that of the relative balance between conventional and nuclear forces in French defense policy thinking.

For most of the Fourth Republic operations overseas and NATO commitments focused attention on the importance of essentially conventional forces. France's colonial wars and pressure from the United States for rearmament in the early 1950s to counter the large Soviet armies in Eastern Europe meant that the limited resources available were spent, in particular, on attempting to provide the conventional manpower and equipment so badly needed. The army, as it had in the past, remained the most important of the three services. The great debate in French defense circles in this period concentrated on whether the nation's interests would be best served by a mass army of citizen soldiers or a highly trained professional force. The debate was resolved, as it had been traditionally, in favor of a citizen army. Despite the development of an atomic energy program in France from 1946,* little thought was given to the production of nuclear weapons until the midfifties. In the first seven years after the war the only interest French defense planners had in atomic weapons was in the protection that would be needed against such devices on the field of battle. For this purpose the structure and tactics of the army were adapted to the defense against an enemy employing such weapons. Planning was therefore centered on the need for light divisions, battle groups, and the easy dispersion of forces, and little attention was given to how French forces themselves would utilize atomic armaments.

Despite this emphasis on conventional forces, however, the continuing desire to restore France to what was considered its rightful position in the world in large part contributed to a growing campaign throughout the 1950s, in particular by a small group within the Commissariat à l'Energie Atomique (CEA), for the development of French nuclear weapons. The main emphasis of the CEA, from its inception by de Gaulle in 1946, was to create a nuclear establishment for essentially peaceful purposes. Increasingly from 1954, however, pressure from a small group of dedicated administrator-technocrats, politicians, and military officers in the CEA pushed successive governments to produce nuclear weapons.[25] Partly as a result the question of a French nuclear program for military purposes was first raised at cabinet level on 26 December 1954 when an interministerial committee was convened by

*The nuclear program was initiated by an ordinance issued by the provisional government establishing the Commissariat à l'Energie Atomique (CEA) on 18 October 1945.

Premier Mendès-France to study the question of the military applications of atomic energy. With the rather inconclusive results of the committee, pressure on the government was increased, especially after the humiliations of Indochina and the Suez affair, by public campaigns, particularly by men like Colonel Ailleret and General Ely.

Colonel Charles Ailleret, who held the post of *commandant des armes speciales* (commandant of special forces), conducted a number of studies on atomic weapons with a small group of officers from 1952 onward. During the 1950s Ailleret was to become one of the major military proponents for nuclear weapons. He conducted his campaign in numerous articles, arguing his case not from firm notions of nuclear strategy but rather on the basis of a narrow view of their military advantages as the most modern weapons. His main contention was that atomic weapons were both cheaper than conventional explosives and also far more effective. Although he had no clear thesis about their strategic and political uses, he did anticipate de Gaulle's later views in his recognition that nuclear weapons would allow France "to assert herself with much more weight in alliances."[26]

Another influential advocate of French nuclear development was General Paul Ely, who returned to France in 1956 from Indochina to become *chef d'état–major-general des armées* (army chief of staff).[27] In this capacity he established a committee to study nuclear weapons and their strategic implications. On the basis of these studies General Ely was the first to advocate the idea of a *force de frappe* (strike force), which was to become so famous under de Gaulle. To Ely, however, the notion of a *force de frappe* was a broad military concept that would include both conventional and nuclear forces. In an article he published in 1957 he argued that France should have the capability to operate anywhere in the world to meet different kinds of threats at different kinds of level.[28] To achieve this flexible response, Ely argued that France required a spectrum of forces (including atomic weapons) to meet varying kinds of conflict. As far as he was concerned, however, such a force had to be placed "fully within the framework of NATO."[29]

The gathering momentum of these pressures on successive governments to initiate a military nuclear program finally led to the decision by Defense Minister Bourges-Manoury to create a nuclear force in the spring of 1957. This was followed by the announcement in April 1958 by Premier Félix Gaillard that 1960 had been set as the target date for the first French atomic test. The decision therefore to create a nuclear force was taken by a government of the Fourth Republic. With the Algerian crisis, however, it was left to the leaders of the Fifth Republic to execute the policy laid down by their predecessors. On assuming power de Gaulle immediately accelerated the atomic weapons program and in 1959 unveiled his plans for a *force de frappe,* which he envisaged as an "advanced instrument representing an enormous potential of massive destruction capable of instant use anywhere on the globe without preliminary mobilization."[30]

Despite the relatively greater emphasis on nuclear technology in the first years of the Fifth Republic, however, the continuation of the Algerian conflict meant that French defense policy had to retain a balance between conventional and nuclear forces. Reflecting this need the minister of national defense declared in 1958 that the possession of nuclear weapons and defense against subversion had equal priority. Thus in his first five-year plan *(loi de programmation)* for defense, between 1960 and 1965, de Gaulle attempted a complete reorganization of national defense that would secure the necessary balance between conventional and nuclear forces.* The French armed forces were organized in three tiers: the Force Nucléaire Stratégique (FNS); the Forces d'Intervention, renamed the Forces de Manoeuvre in 1964; and the Défense Operationnelle du Territoire (DOT). According to the first *loi de programmation,* the Forces de Manoeuvre were designed for the defense of French frontiers, for NATO operations, and also for the defense of the French Community and French interests overseas (intervention outre-mer). It was envisaged that they would consist of five mechanized divisions and one light division together with the necessary naval and air forces. The DOT was to consist of 100 regiments to be used essentially for internal defense.

With the end of the Algerian War and the increasing economic strains facing France, the attempt to create such a wide spectrum of military forces became more difficult. De Gaulle's dilemma was that to reestablish herself as a world power France had to possess both the conventional forces to intervene overseas and the atomic forces to gain membership of the nuclear club. Economic realities, however, dictated that France could not have both, at least on the ambitious scale laid down in 1960. In such circumstances, de Gaulle had little doubt that French objectives of independence and grandeur could best be served through the continuation of the nuclear program, even if this meant scaling down conventional forces.

Although the government claimed with some justification that its defense plans in 1960[31] were an attempt to achieve a balance between conventional and nuclear forces, by the start of the second *loi de programmation* in 1965 a growing imbalance was in fact clearly emerging. This second five-year plan envisaged only three fully armored divisions equipped with AMX tanks and two equipped with older hardware for the Forces de Manoeuvre by the end of the decade. It also reduced the proposed size of the DOT forces from 100 regiments to 25 regiments and one alpine brigade. In terms of resource allocation the declining impor-

*There was, however, an increased emphasis on nuclear weapons in the *loi-programme,* and the new policy had to survive two censure motions in the National Assembly before coming into effect.

tance of conventional forces was clearly demonstrated by the significant reduction in spending on conventional equipment.

Under the second *loi de programmation,* 377 million francs disappeared from the army estimates. There was also a projected reduction of manpower in the armed forces from 1,009,000 in January 1963 to the goal of 700,000 during the late sixties. The ground forces in particular were to be cut back, and national service was gradually to become more selective, thus reflecting the radical change from a mass-citizen army to a highly trained professional force.*

It had now become obvious that the FNS was very much the central feature of French defense policy and that other elements would be sacrificed if necessary to retain its position. From 1965 onward conventional forces in French defense policy therefore were viewed increasingly in the role of support for, and defense of, the FNS. In his study of French defense policy, Mendl argues that after 1970 the main mission of the navy was to protect the nuclear submarines, that of the air defense was to cover the strategic nuclear force, and that of the Forces de Manoeuvre and the forces of the interior was to defend the conventional powered ballistic missile submarines (SSBS).[32] The result of this emphasis was a growing deterioration in French conventional forces in the early 1970s.

The need to eradicate some of the main weaknesses of French conventional forces was recognized by the Pompidou administration in the third *loi de programmation* (1971–75) and by President Giscard d'Estaing in his defense review undertaken in the summer of 1974.[33] It was not, however, until 1976 that concrete improvements were in fact proposed. Speaking to a television audience in May 1976, President Giscard d'Estaing clearly admitted that after fifteen years' preference for nuclear arms there was now a need to help the land, sea, and air forces to catch up in conventional modern arms and equipment.[34] Apparently with this objective in mind defense spending for the period from 1977 to 1982 was projected to increase from its existing level of 17 percent of the budget to 20 percent by 1982. The priority was to be given to the army, which would be reorganized, better equipped, and better trained. Improvements were also to take place in the navy to match its expanded role, which now included sea control and the projection of military power overseas; and the air force was also to be equipped with new aircraft including the Mirage 2000 by the later 1980s.

The 1977–82 *loi de programmation* was seen by many as the beginning of a comprehensive program of conventional improvements intended to implement the new strategy of *sanctuarisation élargie* (extended sanctu-

*The National Service Law of 9 July 1965 broke the institution of universal military service that had lasted since the "Great Revolution."

ary). Upgraded and improved general-purpose forces were necessary, it was argued, to provide the greater flexibility of response required by French defense planners. Some analysts, however, have suggested that careful scrutiny of the plan and what actually happened in the first four years of its implementation indicate that only minor improvements in conventional forces were budgeted for, and the improvements that have taken place have not been all that substantial. Alan Ned Sabrosky, writing in 1981, pointed out that

> . . . . despite the relatively greater emphasis to be given to the general purpose forces in the light of France's new strategy and doctrine, only a minimal redistribution of funds among the Services was projected in the program, and only marginal changes actually seem to have occurred to date. It would appear that bureaucratic and budgeting forces combined in this case to make the present and the future more like the past than might have been anticipated by someone contemplating the potential implications of *sanctuarisation élargie* and flexible response for French defense policy.[35]

There seems little doubt that French nuclear forces have continued to be given priority in the allocation of resources to defense in recent years. According to the 1977–82 *loi de programmation* the French nuclear-powered ballistic submarine fleet was to be expanded to five vessels. The first four submarines (SSBNs) were equipped with sixteen M20 missiles (SLBMs) with single warheads capable of a range of three thousand miles. These were to be followed by a fifth submarine with sixteen M4 missiles each with three multiple reentry vehicle (MRV) warheads and with a range of four thousand miles. It was planned to upgrade the first four nuclear submarines by retrofitting them with the new M4 missiles by 1985. (After much debate it was subsequently decided to proceed with a sixth SSBN—*L'Inflexible*—which would become operational in 1985).[36]

Although the submarine fleet remained the linchpin of the French strategic force (FNS), improvements were also envisaged in other parts of the force as well. France's intermediate-range ballistic missile force (IRBMs) of eighteen SSBS S2 missiles with a range of eighteen hundred miles based in silos on the Plateau Albion in southern France was to be updated with the new SSBS S3 missile with a range of three thousand miles by 1982. At the same time it was envisaged that the manned bomber force of thirty-three Mirage IVA aircraft would remain in service until 1985 and would then be replaced by either light mobile (SX) ballistic missiles or by nuclear-armed cruise missiles (depending on the outcome of various studies being undertaken by the defense ministry).

The strategic forces were not the only nuclear systems to be improved. Studies were also taking place on a replacement for the Pluton short-range (seventy-five miles) tactical nuclear missile, which had been deployed with five regiments since 1974. It was planned also to equip the sixty Mirage III and Jaguar aircraft with the AN-52 tactical nuclear bomb that was under development. The same bomb was to be placed on the new Super-Etendard bombers in due course.

This continuing stress on nuclear forces once again suggests that despite the criticism of many Gaullists (including General Gallois), that the effectiveness of the French nuclear deterrent was being undermined, continuity rather than fundamental change has to a large extent characterized the balance between conventional and nuclear forces in French defense policy.[37] Some improvements in the army and navy in particular do seem to have occurred, and official statements have certainly stressed the importance of general-purpose forces to secure French security interests. However, there can be little doubt that defense planners retain an intense belief that "only nuclear weapons can underwrite French Security and insure that other countries will give it [France] its rank in the concert of nations."[38] As has been shown, this is a trend that, in general, has been continued by the Socialist Mitterand government in the first half of the 1980s.

## *Strategic Thinking in France*

The changing balance between conventional and nuclear forces is both a reflection of and reflected in the constant adjustments that have taken place in French strategic thinking in the postwar period.[39] For much of the Fourth Republic the focus of the strategic debate centered on the two preoccupations of European defense and counterinsurgent campaigns overseas. In the European theater, French conventional forces were seen as playing their part in an overall allied strategy. French forces, like those of French allies, were designed to identify Soviet aggression and delay Soviet advances into Western Europe. During the "pause" resulting from conventional operations it was hoped that time would be gained both for negotiations to take place and for the decision to be made, if necessary, to use nuclear weapons.

Besides these European operations, French forces were also seen as having the important role overseas of countering insurgent activity in France's colonies. The perceived importance and the continuing necessity during the 1940s and 1950s of fighting subversive campaigns led to the development in French defense circles of what became known as *la guerre révolutionnaire* doctrine.[40] The experiences both of Indochina and Algeria led to a growing interest among strategic planners in the writings of Mao Tse-tung and the attempt to produce a coherent theory of counterrevolutionary warfare to defeat France's enemies. To many political and military leaders in France the main threat to French, and indeed Western, security emanated from the indirect strategy adopted by the Communist states in Africa, Asia, and Latin America. The belief in the global significance of the campaign in Algeria in particular led to priority being given in French defense policy in the last years of the Fourth Republic and the early years of the Fifth Republic to the concentration of military resources on countering this perceived threat. *Guerre révolutionnaire* therefore in some ways became the dominant theme in French strategic thinking during this period.[41]

Growing disillusionment, however, with the American guarantee to Western Europe and alliance strategy together with the humiliations emanating from the failures of French counterrevolutionary operations led to a gradual reappraisal of strategic thinking.[42] The conventional and counterrevolutionary elements of French strategy began to give way to a new strand in politico-military thinking as the French military nuclear program gathered momentum from the midfifties onward. Defense planners increasingly emphasized the revolutionary impact of nuclear weapons in introducing a new dimension into warfare. The focus of attention gradually moved away from the need to provide large conventional forces capable of fighting in the European theater and subversive campaigns overseas. Instead greater stress was laid on the importance of preventing wars from breaking out. Nuclear deterrence rather than conventional and unconventional defense now gradually emerged to hold the center of the strategic stage in France.

Although French literature on strategic matters was overshadowed in terms of both "breadth and, to a certain extent, sophistication," by the American writings on the subject, a number of books and articles on the French nuclear force began to appear from 1963 and 1964 onward. Pierre Gallois and André Beaufre in particular entered the strategic debate with various works that provided theoretical justification for the national nuclear deterrent.

In his writings General Gallois argued that the establishment of a "balance of terror" between the superpowers meant that the credibility of the U.S. guarantee to defend Europe had been greatly reduced. It was therefore axiomatic that middle powers, such as France, should not rely on military alliances and should possess their own nuclear forces to enhance their national security. To Gallois nuclear proliferation was inevitable, and indeed desirable, since the greater the number of states possessing nuclear weapons, the greater the degree of international stability through deterrence. In contrast to Gallois's rather extreme and dogmatic analysis, André Beaufre's writings have a subtlety and sophistication that have put him in the front rank of strategic thinkers.[43] Beaufre, unlike Gallois, argued that in order to avoid the danger of international instability that national nuclear forces may create, alliances are necessary. To General Beaufre the decision to use nuclear forces must remain in national hands, but they must be "coordinated" within some form of alliance framework.

Similar, but not identical, views were put forward by another leading

French writer on nuclear strategy, Raymond Aron.[44] Like Beaufre, Aron expressed his opposition to the Anglo-Saxon monopoly of nuclear weapons and supported the utility of French ones. Aron, however, was opposed to the notion of a strictly national deterrent and argued instead for some form of European nuclear force. For Aron, as for General Stehlin, a former air force chief of staff, the nationalistic framework in which de Gaulle developed his *force de frappe* was something to regret.[45] To both it would have been far preferable had the president taken steps to promote a European deterrent with French nuclear armaments as the nucleus of the force.

Although not as extreme, de Gaulle's views in many ways had more in common with Gallois than with either Beaufre or Aron. In 1964 de Gaulle spelled out the concept of "proportional deterrence," which has remained in large part at the center of French strategic doctrine ever since. The General maintained that it was not necessary to match the nuclear capabilities of the United States and USSR, for, he argued, "once reaching a certain nuclear capability . . . the proportion of respective means has no absolute value." What was important was to have the capability to survive an enemy first strike and to inflict an unacceptable degree of damage on an aggressor by way of retaliation. "The deterrent exists," de Gaulle said, "provided that one has the means to wound the possible aggressor mortally, that one is very determined to do it and that the aggressor is convinced of it."[46]

In the early days of the development of the French nuclear force, nuclear strategy was based essentially on the notion of massive retaliation. Once serious aggression by an opponent had been identified, the instantaneous and total use of France's national nuclear force would follow against the adversary's cities. With the development of tactical nuclear weapons, French strategic doctrine underwent a slight modification. It was felt that conventional forces backed by the threatened or actual use of tactical nuclear weapons might cause the enemy to pause and consider the consequences of his actions. Despite the delay, however, between the identification of aggression and the use of strategic weapons, it was thought that the pause could only be of a limited duration. France would then be forced to reply with her whole nuclear arsenal.

In essence French strategic thinking was to remain largely predicated on this modified concept of massive retaliation until the changes introduced by General Fourquet in March 1969.[47] In his speech at the Institut des Hautes Etudes de la Défense Nationale, the new French chief of staff of the armed forces went out of his way specifically to reject not only the all-or-nothing doctrine of massive retaliation but also the idea of meeting aggression at all levels with the appropriate force implicit in the American doctrine of flexible response. Fourquet argued instead for a strategy of "graduated deterrence" designed to raise the threshold of nuclear response higher than that of massive retaliation but not as high as that of flexible response.[48] The chief of staff saw graduated deterrence as a way of overcoming the "suicide or surrender" choice of massive retaliation and retaining the element of deterrence he considered had been downgraded in the doctrine of flexible response. Conventional and tactical nuclear weapons were seen as increasing the number of options available to decision makers, and the threat or even the actual use of the latter would act as a potent deterrent to further escalation. According to General Fourquet's analysis, France's limited conventional forces would be used to test the enemy's intentions. If the test proved positive in the sense that a major threat was identified, French tactical nuclear weapons would then be used as a "shot across the bows" to communicate the seriousness of the French government's resolve to fight a nuclear war if necessary. In so doing it was hoped that time would be bought both for the enemy to reconsider the implications of his aggression and also for the decision to use the nation's strategic nuclear forces should it become necessary.

General Fourquet's address also significantly modified French strategic doctrine in another way. During the mid-1960s the movement for greater independence in French defense policy and the attempt to reestablish French status in the world led to the adoption of the strategy of *tous azimuts.* According to this doctrine, which was introduced largely on the initiative of Fourquet's predecessor, General Ailleret, France was to prepare for attack from any quarter. In an article in December 1967, General Ailleret argued that the threat from Russia had virtually disappeared and that it was difficult to foresee the nature of conflicts that would occur in the future and the coalition of states that might be involved. Because of this uncertainty present alliances were unlikely to be of much value. His conclusion was that France should concentrate on building a complete independent defense system, which would include "a significant quantity . . . of megaton ballistic missiles, with a worldwide range." Ailleret argued that such a system would be aimed at all points of the globe rather than directed against any one power in particular. In so doing France would be able "to wield the maximum power afforded by its national resources."[49]

This rather ambitious doctrine, emphasizing French independence and power, foundered on the rocks of the invasion of Czechoslovakia and the domestic problems of 1968. The increased anxiety over Soviet intentions and the economic crisis were reflected in Fourquet's implicit abandonment of the concept of *tous azimuts.* In his speech Fourquet emphasized the need to meet a potential invasion from the East, which would require the coordination of allied armies. Although Fourquet reiterated the traditional arguments against integration and the importance of sovereign control of defense, the trend of his argument suggested that he believed that France would have to collaborate with her

NATO allies in an emergency. It would seem therefore that the renewed perception of the possibility of a future Soviet threat and the need to reallocate resources away from defense to other areas of domestic society forced the realization on French defense planners that the degree of independence inherent in the *tous azimuts* doctrine was overambitious. French strategic doctrine had to be brought back more into line with the realities of France's position in the world.

The changes in French strategic doctrine foreshadowed by General Fourquet's speech in 1969 were produced in more coherent form in 1976 in speeches made by General Méry. As we noted earlier, the notion of *sanctuarisation totale* (total sanctuary) was discarded, and a new concept of *sanctuarisation élargie* (enlarged or extended sanctuary) was introduced in its place. In essence this meant that French forces would be "expected to engage an enemy in areas approaching France—the classic bataille de l'avant [forward battle]—in place of the earlier 'Fortress France.' "[50] Echoing General Fourquet, it also involved a rejection of the massive retaliation doctrine and the acceptance of a graduated deterrence or flexible response strategy similar to that adopted by NATO in 1967. As such, as discussed above, it required France to have a spectrum of strategic nuclear forces, tactical nuclear forces, and conventional forces to enable her to respond to different threats at the appropriate level.

One of the questions that has arisen over the adoption of the notion of *sanctuarisation élargie* is how France would use its tactical nuclear forces in any future conflict in Europe. Traditional policy has been that in the event of an attack in force tactical nuclear weapons would be fired as "warning shots" to convince the enemy of the seriousness of French resolution to defend herself. As a result it was hoped that the "logic of 'proportional deterrence' would become manifest to the adversary, and France would not be invaded." Such a demonstrative use of tactical nuclear weapons remains an important feature of French policy to this day. At the same time, General Fourquet in the late 1960s and President Giscard d'Estaing in the mid-1970s hinted that they might also be used as battlefield weapons as well. In the event of a Soviet invasion it was suggested that tactical nuclear weapons might have a militarily important function of slowing down the invading armies.[51] As such they were seen not simply as being linked to the strategic deterrent, but they had a war-fighting role as well should deterrence fail. Government policy has, however, remained a little ambiguous on whether the deterrent or battlefield role should be stressed. This ambivalence, which has characterized government policy, has also reflected an uncertainty over whether a limited nuclear war could be fought without escalating automatically to the strategic level.

Critics of government policy on tactical nuclear weapons (TNW) in France have ranged across the spectrum from those (like General Gallois) who wonder whether TNWs are necessary to detonate the strategic forces to those (like Colonel Marc Geneste) who argue that much greater emphasis should be given to planning for tactical nuclear war.[52] Colonel Geneste has maintained that "a properly prepared tactical nuclear defence—with shelters, concealment, deception etc.—would have a 'fighting chance' of defeating a Soviet offensive."[53]

The ambivalence of successive governments on the question of whether a tactical nuclear war could in fact be kept limited has been reflected in the deliberations that have taken place on enhanced radiation (ER) weapons, which have been seen by some observers as a useful military response to the tank threat from the Soviet Union. Research on the so-called neutron bomb was undertaken under the d'Estaing government and was continued by the Socialist President François Mitterand when he came to power in May 1981. It was reported in October 1982 that France had finally decided to go ahead with production of neutron weapons after a lengthy process of review by the Mitterand government.[54] At the same time, however, it is also interesting that although the Socialist administration continued funding of the air-launched missile project and the Hades missile (to replace Pluton), expenditure on tactical nuclear weapons was decreased reflecting perhaps "scepticism about the feasibility of conducting a limited nuclear war" in the new government.[55]

## *Commitments and Capabilities*

The changing balance between conventional and nuclear forces reflected in the evolution of French strategic doctrine has quite naturally had a significant impact on the ability of France to perform her defense commitments outside Europe.

During the 1940s and 1950s the relative emphasis on conventional forces enabled French governments to use military force in an attempt to reestablish control over parts of their empire lost during the Second World War. Operations were mounted in Syria, Indochina, Madagascar, and later Algeria to reassert French authority. Even the Suez expedition in the autumn of 1956 can be viewed as part of the French colonial campaigns that dominated French defense policy until 1962. Egypt was regarded in France as the primary foreign base supporting the rebels in Algeria. It was hoped that the successful invasion of Suez would secure the two objectives of cutting off supplies to the rebels and humiliating Nasser, which would result in a significant moral defeat for the nationalists.

The failure of French military forces in many of her colonial campaigns and the humiliation its defeats caused, contributed, in large part, to the movement away from reliance on conventional forces to greater emphasis on nuclear weapons in defense policy. As in Britain, Suez in particular contributed to the switch in priorities in favor of atomic weapons. As one commentator has put it, "Militarily, the experience of Suez did not support the rationale of a weak nuclear force, but politically and emotionally it strengthened its appeal, particularly as France was shaken less by the Russian menace than by a sense of being deserted by her American ally."[56]

Despite the gradual downgrading of conventional forces in the 1960s to maintain the desired preeminence of nuclear forces, France has not correspondingly reduced the scale of its worldwide commitments to bring them into line with its reduced capabilities. To the contrary, the 1972 white paper speaks in traditionally expansive terms:

> To this . . . search for a European balance founded on real commitments defined with precision is added. . . . another orientation marked by our effort of cooperation. It is a matter here to support, in case of emergency by military aid, the independence of certain states of Francophone culture . . . and to participate in safeguarding and in restoring peace in the Mediterranean basin.[57]

In contrast to France's rather modest capabilities, therefore, French aspirations would seem to remain global, based on the continuing vision of France as an independent world power.

In reality the size of French intervention forces in the 1960s and 1970s suggested that France could not be expected to sustain a lengthy or a large-scale operation. The dilemma can be seen clearly from French overseas operations during the last twenty years. Despite the belief that France has a major role to play in black Africa, the only interventions France has been capable of mounting have been strictly limited ones.* In Gabon in 1964, French forces were used in a small "quick fix" action to reinstate the deposed government after a surprise coup d'état.[58] Operations in Chad in 1968 and likewise in the early period of the Pompidou administration revealed both the limited capability of French conventional forces and the degree of public opposition at home that can be expected in similar operations in the future. The use of French forces in Zaire in 1978, although highly successful in many ways, also demonstrated certain military inadequacies in the rapid deployment force. France was only able to undertake the operation (ironically perhaps) with the aid of American air transport and logistic support.

The recognition of the inadequacies of France's intervention led to the development of a new rapid deployment force in the early 1980s. Despite past criticisms by the Socialists of France's role as the gendarme of Francophone Africa, President Mitterand not only sanctioned the creation of a new highly mobile force (FAR) of forty-seven thousand men for operations both within and outside Europe, but he also dispatched forces to Chad in the summer of 1983 to meet an invasion backed by

---

*France has a series of bilateral security accords with a number of black African states.

Libya.[59] Mitterand was also prepared to contribute French troops to the Western multinational force in the Lebanon in 1983 to try to help re-establish some stability in that war-torn country. At the same time, however, it must be said that further cuts in France's conventional forces as a whole in the early 1980s have continued to raise doubts about its material capacity to sustain a prolonged conflict overseas.[60]

## *Conclusion*

One of the most interesting features of French defense policy in the late 1970s and early 1980s has been the lively debate over the question of continuity and change as reflected in the 1977–82 and 1984–88 defense programs.[61] The arguments presented earlier in this chapter suggest that despite the ambiguity of French official pronouncements when Giscard d'Estaing was in office and the criticisms from left and right, in general no fundamental changes in French defense policy seem to have occurred during the late seventies. It is true that the significance of the new strategy of *sanctuarisation élargie* and the greater willingness to cooperate with NATO (and especially West Germany) should not be underestimated. Nevertheless, traditional Gaullist principles have remained at the heart of French defense planning. These principles include the preeminence of France's strategic nuclear forces and the concern to retain total independence in defense decision making.

It is significant that despite its longstanding opposition to the French nuclear weapons program the Socialist administration that came to power under François Mitterand in 1981 showed no hesitation in taking important nuclear force decisions shortly after assuming office. At the Defense Council meeting on 30 October 1981 the president approved the construction of a seventh SSBN for the FNS as well as the development of the SX mobile IRBM to replace the Mirage IVA bombs in the early 1990s.[62] The "absolute priority" given by the Socialist government to the nuclear deterrent was spelled out very plainly in the 1984–88 *loi de programmation* announced in April 1983. The five-year plan proposed cuts in France's conventional forces and a redistribution of resources so that more could be spent on strengthening the strategic nuclear force. Under the new plan, which caused an outburst of protest from French defense chiefs, spending on the nuclear force was to take up to 32 percent of the total equipment budget.[63] Also the French army was to lose 22,000 of its 312,000 men (representing a 7-percent cut); the air force was to lose 3,500 men, and the Navy 3,500.* Nevertheless, it must be

*Warnings that a planned cut of thirty-five thousand men would weaken the structure of the army and diminish morale may have had some impact on the government, which then announced the cut of twenty-two thousand men.

admitted that despite the uncertainty over the role of the new FAR rapid deployment force there is some indication that conventional forces and a contribution to European defense are beginning to figure more prominently in Mitterand defense policy.

President Mitterand on a number of occasions supported the NATO decision taken in December 1979 to deploy cruise and Pershing II missiles in Europe to counteract the growing threat he perceived from Soviet SS-20 missiles. The president's justification for his outspoken support for NATO policy on this question reflected the essential continuity in French policy even under a Socialist administration. In an interview given to the German magazine *Stern* in July 1981 the president argued that France "needed an equilibrium between blocs. Only thus can France remain as free as possible of any responsibility in the defence of her neighbour, the Federal Republic and thus be able to concentrate on the nuclear deterrence for her own territory."[64] As under Giscard d'Estaing, it was clear that despite support for alliance policy France had no intention of rejoining the integrated military organization of the alliance.

The continuity of policy is also seen in the way that defense planners in the 1970s and 1980s have reemphasized French interest in Franco-German defense cooperation and the establishment of a new West European defense pact based on the Western European Union (WEU) concept. These trends, have been developed further by the Mitterand government and become a highly distinctive feature of its foreign and defense policies. Following discussions on security matters between the foreign and defense secretaries in October 1982 speculation arose about possible German participation in France's nuclear strategy.[65] France was reported to be interested in placing some of its nuclear missiles on West German territory and securing some financial assistance from the West Germans for its expensive nuclear program. The French were apparently less interested in allowing the Germans to have any say in the planning of its nuclear strategy. Again in February 1984 it was reported that there had been an extension of defense cooperation between the two countries particularly at the parliamentary level with the initiation of joint meetings between the two countries' defense committees.[66]

At various times since 1963 the attempt to form a Franco-German alliance has been part of a wider interest in developing a European framework for defense based upon the Western European Union. The French interest in revitalising the WEU was brought out clearly in a speech made by Foreign Minister Jobert to the WEU Parliamentary Assembly in November 1973.[67] Jobert proposed that the WEU should become the forum for future discussions in Europe's defense arrangements. Again in July 1974 the new French foreign minister, Sauvagnaragues, in his first speech before the foreign affairs commission of the National Assembly, emphasized the importance of French independence of NATO and by implication his support of the WEU idea. French government officials also apparently sounded out Common Market partners in August 1974 on the need for a so-called West European Defense Council, which would meet regularly and discuss security problems common to all members.[68] These references to the WEU as a suitable framework for Western European military cooperation have been given even greater prominence on a number of occasions since May 1981 when François Mitterand came to power. In a speech to the Institute des Hautes Études de la Défense Nationale in September 1983, the Socialist prime minister, Pierre Mauray, went out of his way to emphasize the continuity of French thinking on the WEU.[69]

As far as the French are concerned the WEU concept would seem to have a number of distinct advantages. It is an entirely European institution, independent of the United States, and groups together the leading member countries of the EEC. The WEU is seen as a much better forum than the Eurogroup within NATO. In this sense once again the French are taking the lead in attempting to create an independent defense organization that would form the core of the larger designs of the new French president for European Union. The WEU concept would also ensure the important function in French foreign and defense policy of coordinating French policies with those of Germany while at the same time retaining French superiority in the military field. The original protocol establishing the WEU placed a major limitation on West German freedom in military affairs through the German agreement not to manufacture nuclear weapons.[70]

Contemporary interest in the WEU suggests that in many respects French policy is still characterized by what David S. Yost has described as "equivocation."[71] Under Giscard d'Estaing and François Mitterand there has been a realistic acknowledgment "both of France's ultimate dependence on the United States for its own security and the importance of cooperating with other West European countries to enhance the security of Europe as a whole. But "at the same time, France still clings to an assertive independence whenever actual or perceived threats to French interests or slights to French ambitions are concerned."[72] This equivocation can be seen by the way Mitterand, in his early days in office, distinctively emphasized his firm support for an alliance with the Reagan administration on most East-West issues while simultaneously demonstrating French independence in his relations with the Third World in Africa, in Central America, and in the Middle East.[73] For many observers this independence reflects traditional French policies that attempt "to play the game both ways": to be part of an alliance with the benefits that provides while not accepting all of the responsibilities of alliance membership.

France's new African policy in particular illustrates the continuing importance in French foreign and defense policies of the notions of independence and greatness. Despite the various modifications forced on French defense planners by the incompatability of overambitious aspirations and the nation's limited resources, France's aim to pursue these two objectives as far as possible still appears to be an important motivating force. As such, although more pragmatic and realistic than in the mid sixties, French policy in defense is likely to be faced by much the same kind of dilemma as it has faced in the past. How much grandeur and how much independence in defense is compatible with domestic stability and national security? As in the past, the degree to which these two objectives are pursued is likely to be determined both by internal and external events.

The French dilemma, however, runs deeper than this. The problem is not just one of the incompatability between the means available and the objectives sought. There is also a fundamental contradiction between the objectives themselves, which is likely to cause increasing problems for French leaders. Although from the French viewpoint the notion of grandeur involves in part national independence, it also involves the creation of a strong political, economic, and particularly military power base. Without such a national power base, "greatness," at least in terms of the other superpowers, would require the pooling of Western European resources as a whole in some form of European union. So far, because of the relatively small incremental steps that have been taken toward European integration, successive French governments to a certain extent have been able to obfuscate the difficulties of achieving "greatness" while at the same time remaining independent. If the more important decisions of European integration are ever made, the French will have to decide whether to lose a large part of their sovereignty in the hope of achieving "greatness" as part of a European bloc or to retain their national independence and forgo the opportunity of grandeur that membership of a European union could bring.

The problem, as Alan Ned Sabrosky has aptly pointed out, is that France has "not yet reconciled itself to its present position in the world, nor has it yet been willing to focus its efforts entirely on a specific role in world politics. It still pursues, at different times and on different issues, variants of independent, Atlanticist and European policies intended to allow France to interact with other nations while postponing a final resolution of the basic contradictions in its declared interests."[74] The result is an impressive military capability but a defense policy that fails to provide a coherent framework for balancing commitments and capabilities.

## NOTES

1. "In the world as it is," de Gaulle said, "people sometimes feign surprise over the so-called changes and detours of France's actions. And there are even those who have spoken of Machiavellianism. Well, I believe that while the circumstances are changing around us,

in truth there is nothing more constant than France's policy. For this policy, throughout the extremely varied vicissitudes of our times and our world—this policy's essential goal is that France be and remain an independent nation." Quoted by Roy C. Macridis in "French Foreign Policy," in Macridis, ed., *Foreign Policy in World Politics* (Englewood Cliffs, N.J.: Prentice-Hall, 1967), 74.

2. In his study of French military policy Wolf Mendl describes "defense policy as the function of foreign policy." See W. Mendl, *Deterrence and Persuasion: French Nuclear Armament in the Context of National Policy 1945–69* (London: Faber and Faber, 1970), 18.

3. W. L. Kohl, *French Nuclear Diplomacy* (Princeton, N.J.: Princeton University Press, 1971), 6.

4. Ibid., 6–7.

5. *Notes et Etudes Documentaires,* no. 3343 (6 December 1966) (the English translation is in *Survival* [January 1968]: 12–16; and Mendl, *Deterrence and Persuasion.*

6. Mendl argues that "the possibility of American assistance was continually evoked in the years right up to and immediately after the first atomic explosion." Ibid., 57.

7. Ibid., 33 and 31–35.

8. Kohl, *French Nuclear Diplomacy* 133.

9. Ibid., 268.

10. Mendl, *Deterrence and Persuasion* 70.

11. As Mendl says, "Throughout the period since the end of the war, one might well ask, 'Whom were the French watching on the Rhine—the Russians or the Germans?' " Ibid., 35.

12. See D. Schoenbrum, *The Three Lives of Charles de Gaulle* (New York: Atheneum, 1965), 295–300.

13. For the debate about the different interpretations of de Gaulle's tripartite proposal see Kohl, *French Nuclear Diplomacy,* 74–81.

14. De Gaulle, *Mémoires d'espoir: Le Renouveau, 1958–62* (Paris: Plon, 1970), 214–15.

15. Although the French subsequently encouraged the belief that the U.S. did not reply to the 1958 memorandum, Eisenhower did in fact respond on 20 October 1958. See footnote 51 in Kohl, *French Nuclear Diplomacy,* 73.

16. Charles de Gaulle, speech at Centre des Hautes Etudes Militaires, 3 November 1959, 1.

17. "Emploi des différents systèmes de forces dans le cadre de la stratégie de dissuasion," *Revue de défense nationale* (May 1969): 757–67; a translation appears in *Survival* (July 1969): 206–11.

18. See David S. Yost, "The French Defence Debate," *Survival* (Jan.–Feb. 1981): 19–28.

19. *Rapport sur la programmation des dépenses militaires et des équipements des forces armées pour la période 1977–1983,* 8.

20. Reprinted in *Défense Nationale* (June 1976) and *Survival* (Sept.–Oct. 1976). See also J. Baylis, "French Defence Policy: Continuity or Change?" *Journal of the Royal United Services Institute for Defence Studies* (RUSI), 122, no. 1 (March 1977).

21. President's remarks reprinted in *Survival* (Sept.–October 1976).

22. Jean Klein, "La France, l'arme nucléaire, et la défense de l'Europe," *Politique étrangère* (December 1979): 465.

23. See Alan Ned Sabrosky, "The Defense Policy of France," in Douglas J. Murray and Paul R. Viotti, eds., *The Defense Policies of Nations: A Comparative Study* (Baltimore: Johns Hopkins University Press, 1982), 232; and David S. Yost, "French Defense Budgeting: Persistent Constraints and Future Prospects," manuscript presented at the biennial meeting of the Section on Military Studies of the International Studies Association, Kiawah Island, S.C., 8–10 November 1978.

24. David S. Yost, "The French Defence Debate," 21.

25. For a discussion of the effectiveness of this pressure group see Kohl, *French Nuclear Diplomacy* 7.

26. For a discussion of Ailleret's role in the campaign for nuclear weapons see Kohl,

ibid., 21–22, 30–32; Charles Ailleret, *L'Aventure atomique française* (Paris: Grasset, 1968), 114, quoted in Kohl, ibid., 31.

27. Kohl, ibid., 44–47.

28. Gen. Paul Ely, "Notre politique militaire," *Revue de défense nationale* (July 1957): 1040–47.

29. Ibid.

30. General Lavand, quoted in *Le Monde*, 18 August 1960.

31. For a criticism of the balance of conventional forces and nuclear weapons even in 1960 see Bosquet's article in *L'Express*, 25 February 1960.

32. Mendl, "French Defence Policy," *Survival* (April 1968). Mendl also quotes a Gaullist deputy, Jean-Paul Palewski, who in the debate on the military budget for 1968 claimed that "by 1980 the number of squadrons, including transport aircraft, at the disposal of the air force will equal those of the Swiss air force today."

33. See J. Baylis, "French Defence Policy," 40.

34. Speech quoted in *The London Times*, 7 May 1976.

35. Sabrosky, "The Defense Policy of France," 247–48.

36. See "French Foreign and Defence Policies," *Strategic Survey 1981–82*.

37. See Yost, "The French Defence Debate."

38. Méry, "Une Armée pour quoi faire et comment?" *Défense nationale* (June 1976): 13–14

39. For an excellent analysis of French strategic thinking, see Kolodziej, "French Military Doctrine," in F. B. Horton, A. C. Rogerson, and E. L. Warner (ed.), *Comparative Defense Policy*, (Baltimore: Johns Hopkins University Press, 1974).

40. See P. Paret, *French Revolutionary Warfare* (London: Pall Mall Press, 1964).

41. Mendl argues that "from 1954 onwards soldiers and politicians preached the new doctrine of war in which the West, conceived in terms of civilization, defended itself in a constant struggle against the forces of world communism." 101.

42. See especially the two essays by Gallois in *Pour et contre la force de frappe* (Paris: Editions John Didier, 1963) entitled "Chaque puissance nucléaire a deux visages" and "Pierrelatte a ses raisons." See also "The Raison d'Etre of French Defence Policy," *International Affairs* (Oct. 1963): 497–510; and Kohl, *French Nuclear Diplomacy*, 172.

43. See Beaufre's books *Deterrence and Strategy* (London: Faber, 1965) and *Introduction to Strategy* (London: Faber, 1965). See also *Nato and Europe* (New York: Knopf, 1966).

44. Many of Aron's articles about the French force and NATO were written for *Le Figaro*. See, for example, "A l'ombre de l'apocalypse nucléaire," 17 August 1966, and "Force nucléaire nationale et alliance atlantique." See also Kohl, *French Nuclear Diplomacy*, 175.

45. See General Stehlin, "The Evolution of Western Defence," *Foreign Affairs* (Oct. 1963): 70–83, and "French Thoughts on the Alliance," NATO, *Fifteen Nations* (Aug.–Sept. 1964).

46. Press conference, 23 July 1964. *Speeches and Press Conferences*, no. 208 (France: Ambassade de France), 9. See also Gallois, *Balance of Terror* (Boston: Houghton Mifflin, 1961).

47. See *Survival*, July 1969.

48. See Kohl, *French Nuclear Diplomacy*, 162–64, 265–66.

49. See Charles Ailleret, "Défense 'dirigée' ou défense 'tous azimuts,'" *Revue de défense nationale* (December 1967): 1923–32. For English translation see *Survival* (February 1968): 38–43. Despite his dim view of alliances, Ailleret did not go as far as Gallois and argue that all alliances were obsolete (see Gallois, *Balance of Terror*). Quotes are from Ailleret, "Défense 'dirigée' ou défense 'tous azimuts.'"

50. Sabrosky, "The Defense Policy of France," 237.

51. David S. Yost, "The French Defence Debate," 20. The use of French nuclear weapons on German soil has been an issue that German planners have constantly wanted to raise in the renewed Franco-German defense talks in the late 1970s and early 1980s—with little success. See *Der Spiegel*, 21 February 1984.

52. See Gallois, "French Defense Planning—The Future in the Past," *International Security* (Fall 1976); and "The Future of France's Force de Dissuasion," *Strategic Review*

(Summer 1979). For a discussion of Col. Marc Geneste's ideas, see Yost, "The French Defence Debate," 24.

53. Yost, "The French Defense Debate," 24.

54. *International Herald Tribune,* 15 October 1982.

55. See *Strategic Survey 1981–82,* 64. For a discussion of the ambiguity of French nuclear policy under Mitterand see Jolyon Howorth, "Consensus of Silence: The French Socialist Party and Defence Policy under François Mitterand, *International Affairs* (Autumn 1984).

56. Mendl, *Deterrence and Persuasion* p. 106.

57. *French White Paper on National Defense,* 1972 (Ambassade de France, Service de Presse et d'Information).

58. The problems of such "quick fix" actions are dealt with in Kolodziej's chapter, "French Military Doctrine," in F. B. Horton, A. C. Rogerson and E. L. Warner, eds., *Comparative Defense Policy* (Baltimore: Johns Hopkins University Press, 1974).

59. See *London Times,* 19 August 1983. For a discussion of the ambiguity of Mitterand's policy on the FAR see Howorth, "Consensus of Silence."

60. Kolodziej, 45. France has shown a reluctance to become too heavily involved in the fighting in Chad despite the new rapid deployment force.

61. See Baylis, "French Defense Policy," and Yost, "The French Defence Debate."

62. *Strategic Survey 1981–82.*

63. In a letter leaked to the French daily newspapers the French army Chief General Jean Delauney warned the government in late 1982 that the cuts would significantly weaken the army. He was replaced as chief of staff in March 1983. See *Financial Times,* 7 December 1982 and *London Times,* 10 March 1983. The emphasis on the nuclear deterrent has also resulted in criticisms of the nuclear deterrent itself in terms of its cost, independence, and effectiveness. See *Guardian,* 17 February 1984. Much the same criticisms about the neglect of France's conventional forces were made by General Lacaze, the chief of staff before his retirement in July 1985. See *London Times,* 30 July 1985.

64. *Strategic Survey, 1981–82,* 62.

65. *London Times,* 8 March 1984.

66. Ibid. The mid-1980s saw particular attention being given to improving France's capabilities to fight in West Germany from the start of any war. See *International Herald Tribune,* 3 July 1985.

67. *Evening Standard,* 19 August 1974.

68. Ibid.

69. *London Times,* 15 March 1982 and *NATO Review* 31, no. 5. (1983). By June 1984 the French had managed to create a wider basis of interest in the WEU than ever before. Besides West Germany, Britain was also giving cautious support to the WEU concept. See *London Times,* 19 June 1984. For a discussion of the distinctive nature of Mitterand's approach to defense see Howorth, "Consensus of Silence."

70. Mitterand has continued another traditonal theme by encouraging close cooperation with West Germany. Continuous security talks have taken place since October 1982. Reports have suggested that France has been prepared to discuss its nuclear policy with West Germany and has sought financial help from the Federal Republic for its expensive nuclear force. See *London Times,* 22 October 1982 and *Hildersheimer Allgemeine,* 24 February 1984.

71. Yost, "French Defence Budgeting," 35.

72. Sabrosky, "The Defense Policy of France," 232.

73. See *Strategic Survey, 1981–82.*

74. Sabrosky, "The Defense Policy of France," 260.

# Glossary of Strategic Terms

## Prepared by Colin McInnes

ACTIVE DEFENSE. The attempt to intercept and destroy offensive forces before they reach their target. See also PASSIVE DEFENSE.

AIRBURST. Explosion of a nuclear warhead above a target rather than on contact. Used to alter blast and radiation effects.

ANTIBALLISTIC MISSILE (ABM). Missile designed to destroy incoming ballistic missiles and/or reentry vehicles. See also BALLISTIC MISSILE DEFENSE.

AREA DEFENSE. The active defense (q.v.) of a widespread geographic area in which a number of targets might be located, or through which offensive forces might have to pass to reach their targets. See also TERMINAL DEFENSE.

ARMS CONTROL. Restraint on the deployment and/or characteristics of weapons. International agreement is generally implicit. See also DISARMAMENT.

ASSURED DESTRUCTION. The ability to inflict unacceptable damage upon an adversary in any circumstances (including after a surprise first strike [q.v.]). Associated with U.S. second-strike deterrent until the early 1970s, when it acquired the (sometimes pejorative) acronym MAD (mutual assured destruction). See also DAMAGE LIMITATION.

BALANCE OF TERROR. State of equilibrium caused by the possession of forces capable of inflicting unacceptable damage. Frequently associated with second-strike forces (q.v.) when a first strike (q.v.) is deterred (q.v.) by the threat of assured destruction (q.v.).

BALLISTIC MISSILE. Missile with non–air-breathing engine that uses thrust alone for lift. Once thrust has stopped, its trajectory is determined by gravity and aerodynamic drag. See also CRUISE MISSILE.

BALLISTIC MISSILE DEFENSE (BMD). Active defense (q.v.) against strategic ballistic missiles. Deployment was severely limited by SALT I, but general interest was maintained by both superpowers and received new emphasis in the late 1970s and 1980s with technological developments (including American fears of a Soviet breakthrough) and America's requirement for a survivable ICBM basing mode.

BATTLEFIELD NUCLEAR WEAPONS. NATO term for nuclear weapons with ranges up to two hundred kilometers.

BRINKMANSHIP. An attempt to force an opponent to reach agreement by deliberately creating the risk of war or escalation.

$C^3$. Command, Control, and Communications. Also $C^3I$ ($C^3$ and Intelligence). Defined by Paul Bracken as "an arrangement of facilities, personnel, procedures, and means of information acquisition, processing, and dissemination used by a commander in planning, directing, and controlling military operations."

Central strategic systems. American term designating those offensive nuclear weapons it considers essential to maintaining the strategic nuclear balance. See also Strategic nuclear weapons.

Circular error probable. Measure of missile accuracy derived from the radius of a circle centered upon the target and within which 50 percent of warheads fall.

Civil defense. Occasionally active, but more usually passive defense (q.v.) of civilian population, industry, and resources, generally using shelters, evacuation, and dispersal.

Clean weapon. Nuclear weapon in which the residual radiation effects are low. See also Dirty weapon.

Collateral damage/casualties. Damage to people or property caused by a nuclear weapon not directly targeted against them. Implicitly the result of a counterforce strike (q.v.).

Controlled Response. A deliberately limited reply to aggression designed to persuade an opponent to practice restraint and avoid escalation (q.v.). See also Escalation control and Flexible response.

Conventional war. Generally refers to a war below the level of a nuclear conflict, but above that of guerrilla warfare or terrorism. The phrase has also been used by some to refer to nonchemical, nonbiological, and nonbacteriological war.

Counterforce strike. An attack directed against *military* targets (including $C^3$ centers, supply depots, and lines of communication). See also Countervalue.

Countervailing strategy. American strategic doctrine approved in 1980 by the Carter administration's Presidential Directive 59 (P.D. 59). Essentially a refinement of Schlesinger's limited nuclear options (q.v.), it sought to establish further flexibility at the strategic nuclear level and present the president with a number of options for retaliation, so that *at a variety of levels of exchange* Soviet aggression would either be defeated or incur unacceptable costs.

Countervalue strike. An attack directed against the social fabric of a nation (including population, industry, and resources). See also Counterforce.

Coupling (strategic). The linking of one area of conflict to another. Associated particularly with the NATO strategy of "coupling" in Soviet eyes a conventional invasion of Europe with the threat of retaliation by U.S. strategic weapons (q.v.).

Cruise missile. Missile with an air-breathing engine and using aerodynamic surfaces for lift (as with an aircraft). Guidance is by remote control or inertial navigation. See also Ballistic missiles.

Damage limitation. The minimization of the destruction caused by (nuclear) attack through the use of preemptive strikes (q.v.) and active and passive defenses (q.v.). Advanced as a strategic doctrine by U.S. Secretary of Defense McNamara in the early 1960s, but soon abandoned due to cost and political problems. It is now being talked of again. Also used in conventional warfare.

Deterrence. The prevention of military action by the threat of punishment (e.g., unacceptable retaliation) or by denying the potential adversary his goals. See War fighting.

Extended deterrence. Attempt to use deterrent threats to protect a third party (e.g., the U.S. guarantee to Europe).

INTRAWAR DETERRENCE. Attempt to prevent certain actions in a war through the threat of punishment or escalation.

MINIMUM DETERRENCE. Attempt to prevent attack through reliance on a small retaliatory force capable of destroying a limited number of targets.

DIRTY WEAPON. Nuclear weapon in which the residual radiation is high (e.g., neutron bomb). See also CLEAN WEAPON.

DISARMAMENT. The abolition or reduction of armaments. Two types exist: unilateral disarmament, whereby one power reduces its level of armaments without a reciprocal reduction from opposing power(s), though one may be hoped for; and multilateral, whereby opposing powers conduct reciprocal reductions. See also ARMS CONTROL.

ESCALATION. An increase in the scope and/or intensity of conflict.

ESCALATION CONTROL. Concept associated with limited nuclear options (q.v.) whereby decision makers can escalate at the strategic nuclear level in a deliberate and controlled manner, allowing time for the enemy to consider the implications of its action.

ESCALATION DOMINANCE. The maintenance of military superiority at each level of a confrontation. Regarded by some as essential to deterrence (q.v.), by others as impractical.

HORIZONTAL ESCALATION. A shifting or widening of response away from the specific area of aggression to territory suitable for a defensive response. Associated with the Reagan administration, and particularly its Persian Gulf and Middle East security policies. Also called "geographic escalation" or "war-widening strategy."

ESSENTIAL EQUIVALENCE. Strategic nuclear balance in which Soviet advantages in certain areas (e.g., throw weight and numbers) are offset by American advantages in others (e.g., MIRVing and accuracy). Although individual force characteristics may therefore be asymmetrical, there is rough equality and therefore strategic stability. Originally coined by Secretary of Defense Schlesinger, the term was extensively used by the Carter administration in SALT II.

FAIL-SAFE. Mechanism to prevent the accidental or unauthorized use of nuclear weapons.

FIRST STRIKE. Initial offensive move in a war. Often implies a nuclear counterforce strike (q.v.). See also SECOND STRIKE.

FIRST USE. The initial use of nuclear weapons. In the NATO context it frequently refers to the deliberate escalation (q.v.) from conventional (q.v.) to tactical nuclear war (q.v.) in accordance with the doctrine of flexible response (q.v.).

NO FIRST USE. Policy or agreement not to use nuclear weapons unless one's oponent has already escalated to that level.

FLEXIBLE RESPONSE. The NATO strategy that involves meeting aggression with direct defense at the same level, but escalating if the aggression cannot be contained at that level. Originally outlined by U.S. Secretary of Defense McNamara in the early 1960s, a version of which was accepted by NATO in 1967 after protracted debate. See also MASSIVE RETALIATION.

FORWARD DEFENSE. NATO commitment since 1957 to defend its territory as far to the east as possible. Military ideas for changing this strategy have been met by political pressure from West Germany.

FRATRICIDE. Untested defensive tactic whereby nuclear explosions from incoming warheads create such local turbulence that other incoming missiles are rendered inoperative. This formed the basis of President Reagan's 1982 Closely Based Spacing mode ("dense pack") for MX ICBM deployment, rejected by Congress. See also PIN-DOWN STRIKE.

GENERAL NUCLEAR RESPONSE. The widespread use of offensive nuclear weapons in retaliation to nuclear attack. See also LIMITED NUCLEAR OPTIONS.

HARDENING. The use of specially strengthened materials to reduce vulnerability to (nuclear) attack.

INTERDICTION STRIKE. Air strikes behind an enemy's front line targeted against $C^3$ facilities and supply lines.

INTERMEDIATE NUCLEAR FORCES (INF). 1980s designation for long-range theater nuclear forces (q.v.) e.g., ground-launched cruise missiles (GLCMs).

LIMITED NUCLEAR OPTIONS. U.S. strategic nuclear targeting doctrine enunciated by Secretary of Defense Schlesinger in January 1974 emphasizing flexibility and selectivity in the employment of nuclear weapons. The NCA (q.v.) would possess a variety of options, including limited strikes by small "packages" of nuclear weapons, thus avoiding the choice between surrender and massive retaliation (q.v.).

LIMITED WAR. A war limited in its geographic scope, military scale, and/or weapons used.

LIMITED NUCLEAR WAR. Nuclear war in which voluntary restraint is exercised not to escalate beyond a certain level or geographic area.

MASSIVE RETALIATION. Nuclear posture associated with Secretary of State John F. Dulles after the Korean War. In its most extreme form, conventional forces deployed in forward defense (q.v.) would detect aggression and act as a "trip-wire" for an all-out nuclear response against the aggressor's heartland. Dulles's public statements, however, indicate that he envisaged a rather more flexible policy than this.

NATIONAL COMMAND AUTHORITY (NCA). U.S. chain of command defining the person with sole authorization for the use of nuclear weapons (the president if he were still alive, then vice president, secretary of defense, etc.).

NUCLEAR WEAPONS FREE ZONE. Geographic area where nuclear weapons have been banned by international agreement. Recent interest has focused on the Balkans and Nordic Europe but with little success.

PARITY (STRATEGIC NUCLEAR). American term used particularly in SALT I referring to rough numerical equality in central strategic systems (q.v.). See also ESSENTIAL EQUIVALENCE.

PASSIVE DEFENSE. The attempt to protect targets through hardening (q.v.), dispersal, or removal from the reach of offensive forces. See also ACTIVE DEFENSE.

PIN-DOWN STRIKE. Limited offensive strike designed to prevent the launch of opposing ICBMs by the local turbulence caused by nuclear explosions. See also FRATRICIDE.

PREEMPTIVE STRIKE. An attack launched in anticipation of an opponent's attack and intended to minimize the effects of the opponent's action.

PRELAUNCH SURVIVABILITY. The ability of a weapons system to survive enemy attack before being launched.

PROLIFERATION (NUCLEAR). The spread of nuclear weapons.

VERTICAL PROLIFERATION is the quantitative increase in existing arsenals.

HORIZONTAL PROLIFERATION is the acquisition of nuclear weapons by states without such arsenals.

QUICK-REACTION ALERT. A condition when forces are kept in a high state of readiness so that they can instantly disperse or launch attacks.

RAPID DEPLOYMENT FORCE (RDF). Conventional force initially proposed by the Carter administration with a quick reaction and intervention capability for the defense of vital Western interests in the Third World, particularly the Persian Gulf.

SECOND-STRIKE CAPABILITY. The ability to survive a nuclear first strike (q.v.) and still retaliate.

SELECTIVE EMPLOYMENT OPTIONS. The limited use of NATO INF (q.v.) to signal a willingness to escalate (q.v.) to general nuclear war.

SINGLE INTEGRATED OPERATIONAL PLAN (SIOP). U.S. plan periodically revised for the targeting of its central strategic forces (q.v.). It was originally drawn up in 1960 to avoid duplication of targets by different commands.

STRATEGIC TRIAD. The three basic launch modes for strategic nuclear weapons (q.v.): ICBMs; nuclear missile–carrying submarines; and long-range nuclear bombers. Maintained by the U.S., USSR, and arguably France and the PRC. Justified by the U.S. as complicating a disarming first-strike (q.v.) by an enemy and guarding against the unexpected vulnerability of any one "leg."

STRATEGIC WEAPONS (NUCLEAR). The definition of strategic nuclear weapons is currently unclear. Previously they were considered those long-range weapons targeted against an enemy's heartland, while tactical nuclear weapons had a shorter range and greater accuracy for use on the battlefield. This distinction has been consistently eroded (particularly through the emergence of long-range theater weapons).

SUFFICIENCY (STRATEGIC NUCLEAR). Doctrine introduced by Nixon in 1970 involving an assured destruction second-strike capability (q.v.); controlled responses (q.v.) to deter limited nuclear attacks; and no attempt to achieve a first-strike capability (q.v.). Essentially a modification to flexible response (q.v.), it was only with limited nuclear options (q.v.) that the required increase in strategic flexibility was achieved in operational planning.

TACTICAL NUCLEAR WEAPONS. See STRATEGIC WEAPONS.

TERMINAL DEFENSE. The active defense (q.v.) of a single target or closely spaced group of targets. See also AREA DEFENSE.

TERRITORIAL DEFENSE. Conventional defense (q.v.) of homeland with negligible offensive potential. Seeks to deter through a strong defensive posture rather than the threat of retaliation.

THEATER NUCLEAR WEAPONS. Nuclear weapons limited to one theater of operation (e.g., Europe). See also INF.

WAR FIGHTING. A posture designed to deter (q.v.) by the threat of being able to fight and win a nuclear war. It is therefore associated with deterrence by denial rather than by punishment. The idea received increased interest in the West during the Carter and particularly the Reagan administrations, and is associated with ideas of a protracted nuclear war. It has generally been seen as the basis for Soviet military doctrine. See also DETERRENCE.

WITHHOLDING STRATEGY. American concept of preserving enemy targets from nuclear attack to allow intrawar deterrence (q.v.) and intrawar bargaining. First introduced by the Nixon administration in 1974. Associated with escalation control (q.v.).

# Abbreviations

| | |
|---|---|
| AAM | air-to-air missile |
| ABM | anti-ballistic missile |
| ADM | atomic demolition mine |
| AEW | airborne early warning |
| AMF | Allied Mobile Force |
| ANZAM | Australia, New Zealand, and Malaya |
| ANF | Atlantic Nuclear Force |
| ARVN | Army of (South) Vietnam |
| ASEAN | Association of South-East Asian Nations |
| ASM | air-to-surface missiles |
| ASROC | anti-submarine rocket |
| ASW | anti-submarine warfare |
| BAOR | British Army of the Rhine |
| BMD | ballistic missile defence |
| CENTO | Central Treaty Organization |
| CEA | Commissariat a l'Energie Atomique |
| CEP | circular error probable |
| COMECON | Council for Mutual Economic Aid |
| CPSU | Communist Party of the Soviet Union |
| CVA | attack aircraft carrier |
| CVAN | nuclear-power attack carrier |
| DOT | Defense Operationnelle du Territoire |
| ECM | electronic counter-measures |
| EDC | European Defence Community |
| EEC | European Economic Community |
| ESC | European Security Conference |
| FBS | forward-based nuclear delivery systems |
| FNS | Force Nucléaire Stratégique |
| FOBS | fractional orbital bombardment system |
| GCD | general and complete disarmament |
| GVN | government of (South) Vietnam |
| IAEA | International Atomic Energy Agency |
| ICBM | intercontinental ballistic missile |
| IRBM | intermediate-range ballistic missile |
| KT | kiloton |
| MAD | Mutual Assured Destruction |

| | |
|---|---|
| MARV | Maneuverable Re-entry Vehicle |
| MBFR | mutual balanced force reductions |
| MIRV | multiple independently targetable reentry vehicle |
| MLF | Multilateral Force |
| MRBM | medium-range ballistic missile |
| MRCA | multiple-role combat aircraft |
| MRV | multiple reentry vehicle |
| MT | megaton |
| NATO | North Atlantic Treaty Organization |
| NLF | National Liberation Front (of South Vietnam) |
| NPT | Non-Proliferation Treaty |
| OTH | over-the-horizon-radar |
| PLA | People's Liberation Army |
| RV | reentry vehicle |
| SAC | Strategic Air Command |
| SALT | Strategic Arms Limitation Talks |
| SEATO | South-East Asia Treaty Organization |
| SLBM | submarine-launched ballistic missile |
| SRAM | short-range attack missile |
| SSBN | nuclear-powered, missile-firing submarine |
| STOL | short take-off and landing |
| TNT | Tri-nitro-toluene |
| ULMS | undersea long-range missile system |
| VLF | very low frequency |
| VTOL | vertical take-off and landing |
| WEU | Western European Union |

# Index